FREGE AND THE PHILOSOPHY OF MATHEMATICS

FREGE AND THE PHILOSOPHY OF MATHEMATICS

Michael D. Resnik

Cornell University Press

Ithaca and London

Copyright © 1980 by Cornell University

All rights reserved. Except for brief quotations in a review, this book, or parts thereof, must not be produced in any form without permission in writing from the publisher. For information address Cornell University Press, 124 Roberts Place, Ithaca, New York 14850.

First published 1980 by Cornell University Press.
Published in the United Kingdom by Cornell University Press Ltd., 2-4 Brook Street, London W1Y 1AA.

International Standard Book Number 0-8014-1293-5
Library of Congress Catalog Card Number 80-11120
Printed in the United States of America
Librarians: Library of Congress cataloging information appears on the last page of the book.

TO JANET

in appreciation of her
courage and beauty

Contents

CONTENTS

[8]

Acknowledgments

The fact that this book was written over a period of four and one-half years, although perhaps not leading to much improvement in my own contributions, did afford me the opportunity to obtain very useful criticisms and suggestions from a number of students and scholars. I wish to acknowledge them now.

The idea of organizing a discussion of contemporary philosophy of mathematics around the more historical themes of the debates between Frege and his contemporaries was inspired by the Ph.D. dissertation on mathematical truth written by one of my students, Anthony Coyne. My intellectual debts to W. V. Quine are considerable, as will be evident in the text. I have also profited from the provocation furnished by Paul Benacerraf's two papers "What Numbers Could Not Be" and "Mathematical Truth" and from the intellectual courage shown by Mark Steiner in his book *Mathematical Knowledge,* where he takes mathematical practice much more seriously than did most earlier philosophers.

I am grateful for the encouragement and voluminous correspondence I have received from Philip Kitcher. He sent me extensive comments on an earlier draft. The written remarks of Charles Chihara, Henry Kyburg, Rolf Eberle, Mark Steiner, W. V. Quine, John Corcoran, Vann McGee, John McLean, and several anonymous publishers' referees have had their impact on the book. So have conversations with James and Catherine Anderson, Richard Grandy, Willard Mittleman, Richard Nunan, Jay Rosenberg, Frederick Schmitt, Brian Skyrms, and Michael Smith.

ACKNOWLEDGMENTS

Dwayne Walls provided editorial assistance and Jayne Fleener, Wendy Lotz, and Willard Mittleman helped with technical details in the final version. Carolyn Shearin patiently typed and typed and typed; without her assistance there would have been no book. William Hobbes helped with the typing earlier, and Claire Miller stepped in, whenever needed, to do whatever had to be done. Her biggest contribution has been to lighten the administrative load of my first term as chairman of the Philosophy Department of the University of North Carolina at Chapel Hill—during which I have done most of the writing. I am grateful to David Falk, who assisted by giving me a leave during the spring of 1975, and to the Smith Fund of the University of North Carolina, which paid for some of the research expenses.

I am grateful to Universitetsforlaget, publisher of *Inquiry,* for permission to use material now appearing in Chapter Five from my article "Frege as Idealist and then Realist," which appeared in Volume 22, No. 3 (Autumn 1979). I also thank the editors of *Philosophy and Phenomenological Research* for allowing me to make use of my article "The Frege-Hilbert Controversy," which appeared in Volume 34, No. 3 (March 1974), of that journal and is now a part of Chapter Three.

I also wish to thank Bernhard Kendler and Marilyn Sale of Cornell University Press for their cooperation, encouragement, and assistance. I am grateful to John Barger for his help with the proofreading.

My apologies to those who helped but have been forgotten.

MICHAEL D. RESNIK

Chapel Hill, North Carolina

FREGE AND THE
PHILOSOPHY OF
MATHEMATICS

Introduction

Frege's Importance

During the last half of the nineteenth century the German philosopher and mathematician Gottlob Frege made revolutionary contributions to logic, the philosophy of mathematics, and the philosophy of language. Unfortunately, Frege's contemporaries failed to understand his work, and he died a bitter and unappreciated man. Only within the last twenty-five years has the general philosophical public come to appreciate and recognize Frege's greatness.

Frege's first achievement came in the field of mathematical logic. In 1879 he published a small book called *Conceptual Notation,* which contained the first formulations of quantificational logic, identity theory, and second-order logic. This work broke an impasse that logic had faced since the time of the Greeks, for it unified truth-functional and quantificational logic while simultaneously extending the latter to handle the logic of relations. This contribution alone would have won Frege a permanent place in the history of conceptual thought. It rescued logic from scholasticism and pedantry and gave it a career as a deep and important branch of mathematics. In this book Frege also defined the ancestral of a relation in terms of second-order logic and, as an application of this definition, derived a generalization of the principle of mathematical induction within second-order logic (his formula (81)).

Inspired by Leibniz's claim that arithmetic could be reduced to logic and by his own success with mathematical induction, Frege next turned to the question of whether arithmetic as a whole can be derived from logic. His research in support of this thesis led to an analysis of counting and to a definition of the natural numbers in terms of equivalence classes of equinumerous classes. Both are widely accepted today, despite the many upheavals that have occurred in the foundations of mathematics since Frege's time. Again taking direction from Leibniz, Frege argued that all mathematical results—and the derivation of arithmetic from logic in particular—must be formulated in a language of impeccable rigor. To make this possible he developed the first modern, formal system for logic.

Frege's informal philosophical analysis of arithmetic is contained in his *Foundations of Arithmetic* of 1884. He augmented this in 1893 and 1903 with the two volumes of his *Basic Laws of Arithmetic,* which were devoted to the formal presentation of his analysis of number, and whose rigorous standards were not matched until the 1920s. Frege also anticipated twentieth-century logicians by clearly distinguishing between the formal system and its interpretation, and by realizing that some sort of proof must be given concerning the relationship between them.

Frege's formal system for arithmetic and set theory was discredited by the discovery in 1901 that it contained the Russell paradox. Although Frege attempted to repair his system, he had no influence upon subsequent developments in axiomatic set theory. Yet it should be mentioned that Frege set the stage for the dramatic history of that subject. For until the paradoxes appeared crystalized in Frege's system, it was all too easy to ignore them.

After the discovery of the paradoxes, some people denigrated Frege's mathematical contributions, but it should be emphasized that most of these contributions are not tied to the context of his particular formal system. For example, his analysis of counting can be reproduced within any formulation of set theory. In addition to his work on arithmetic, Frege made an important methodological study concerning the distinction between axioms and definitions which is contained in his masterful critique of Hilbert's views on implicit definitions.

Frege's great advances in logic were made possible by a new insight into the nature of logical form. Yet he could not have achieved that insight, or made sense of it, without delving into some of the basic issues in the philosophy of language and philosophical logic. His research in this area, presented in a few essays, is far better known and discussed today than is his work in the foundations of mathematics. Indeed, philosophical logic and the philosophy of language is dominated by problems that not only trace back to Frege, but whose very formulations still use his terms. By contrast, his contributions to mathematical logic, although of extreme historical importance, have been so updated and transformed that a student of mathematical logic need never read Frege.

This book focuses on Frege's philosophy of mathematics. But a brief survey of Frege's principal contributions to the philosophy of language is in order. Perhaps the most important of these is his distinction between meaning (sense) and reference, and his correlative account of indirect discourse. His context principle—the dictum that words have meaning only in the context of a sentence—has been heralded by several subsequent philosophers, the most notable being Quine and Wittgenstein. To Frege we owe also the rejection of the subject-predicate-copula analysis of sentences and its replacement by an analysis through general and singular terms. Moreover, he argued persuasively that existence is not an ordinary predicate of individuals, but rather a quantifier, and thus a predicate of predicates. Frege argued extensively against subjective theories of meaning and in favor of the recognition of meanings as mind-independent abstract entities. The influence of his doctrines on the works of Church and Carnap has given a strong Fregean flavor to much contemporary theorizing about meaning. His approach to analyticity was also influential during the first half of this century. Frege defined a statement as analytic when it is one that can be proved through logic and definitions alone. This yielded a far more rigorous characterization of analyticity than any previous philosopher had given, and was a touchstone of the positivistic approach to the a priori.

Let us turn to the main topic of this book: Frege's contributions to the philosophy of mathematics. Here he completely revolutionized the field by demolishing the naive views of his contemporaries and predecessors and by furnishing a model for the type of precision needed in treatments of the philosophy of mathematics. The chief feature of this model is that it attempts to confirm a philosophical view through mathematical results. In Frege's case the model consisted of using a mathematical derivation of the elements of number theory from set-theoretic axioms as part of a demonstration of his philosophical thesis that arithmetic is a branch of logic. Since then, most major philosophical views of mathematics have been supported via mathematical results, and several programs for the foundations of mathematics—such as Hilbert's and Brouwer's—have arisen from philosophical views.

Frege denied Kant's view that arithmetic is synthetic a priori, but he agreed with Kant that geometry is synthetic a priori and based upon intuition. Although most subsequent adherents to the thesis that mathematics is part of logic have not sided with Frege on the last point, they have absorbed another of his views: that numbers are objects. This view is tied to Frege's view of logical form and his theory of the quantifiers. According to Michael Dummett, the view now seems "barely disputable" (Dummett 1973:xx),* but a reading of some of the philosophers who preceded Frege will show how much he advanced our understanding of the status of numbers and numerals.

We also owe to Frege a clear statement of ontological Platonism with regard to mathematics. This is the view that mathematical objects are neither spatiotemporal nor mental, and exist in a mind-independent realm, and that the statements of mathematics are true or false according to whether they describe these objects and their relationships correctly. Frege was also an epistemological Platonist because he held that we know mathematical objects through a type of perception that is analogous to but distinct from ordinary perception of physical objects. Frege's view of mathematics was also methodologically Platonistic, since he used impredicative definitions, the law of the excluded middle, and other principles that subsequent logicians have called into question. As a result, the study of Frege makes a particularly suitable vehicle by which to become acquainted with some of the major contemporary issues in the philosophy of mathematics.

---

*In this book, I refer to a work by enclosing the author's last name in parentheses. If I cite several works by the same author, I use the year of publication to identify the one under current reference. The titles and other information can be found in the Bibliography. Where possible I have tried to cite the original pagination so that the reader will not be dependent upon a particular edition to track down my references.

[*16*]

Frege's Argument and the Structure of This Book

Frege's thesis that arithmetic is a branch of logic can be broken down into three subtheses: first, that the objects of arithmetic are logical objects—a special type of abstract entity; second, that all mathematical concepts can be defined ultimately in terms of concepts that are purely logical; finally, that the theorems and axioms of mathematics can be derived from the basic axioms of logic, using only definitions and the rules of inference of logic itself. Frege used two strategies to support this thesis. On the one hand, he cleared the air through some of the most penetrating polemics in the history of philosophy. On the other, he offered his own positive analysis and technical results.

Frege's foes included some of the most famous philosophers of the last couple of centuries. His archenemy, Mill, offered a physicalistic account of numbers and held that arithmetic is an empirical science based upon induction. While Frege dismissed Mill's view as naive, he found the mentalistic view of arithmetic almost equally repulsive and far more dangerous. This view calls the objectivity of mathematics into question by asserting that the meanings of mathematical terms are the ideas that we associate with them. It can be found in Locke, in Husserl, and in the writings of many of the logicians of Frege's time, who held that the proper business of logic is the psychology and regulation of thinking. Their work, as a very cursory examination will show, could have completely stagnated the field.

Formalism also flourished in Frege's time. This view—that mathematics simply is a game of symbol manipulation—was very popular among nineteenth-century mathematicians and still has adherents among contemporary mathematicians. Late in Frege's career, Hilbert introduced deductivism, according to which mathematics simply deduces consequences from arbitrary axioms that implicitly define the terms they contain and the domain they govern. In this view, consistency turns out to be the sole criterion of mathematical existence and truth. Frege's devastating criticisms of these views brought about their retrenchment and changed the style of the philosophy of mathematics.

Yet philosophical views die hard. Even the views Frege attacked lay dormant, to evolve only later into more sophisticated versions that take into account modern results in the foundation of mathematics. For example, although the early formalists confused use and mention and had no idea of how to construct a formal system, in recent years Curry has developed a formalist philosophy of mathematics which avoids many of Frege's criticisms. Moreover, not only does Hilbert's finitism bear affinities to the early formalists, but some of its important features are due to Hilbert's desire to avoid the pitfalls that Frege pointed out in formalism (Hilbert 1922:165). Recently, Putnam has revived deductivism.

My main purpose in this book is to advance our understanding of Frege's philosophy of mathematics. To this end I have followed Frege's lead and discussed his opponents and his criticisms of them before turning to his positive views. However, I think that our understanding of Frege and of the philosophy of mathematics more generally can be enhanced by examining some modern views that have doctrinal antecedents in theories that Frege attacked. Thus, my treatment includes an examination of the views of Curry, Hilbert, and Putnam mentioned above. I have attempted to show not only that they can be seen as responsive to Frege's criticisms but also that they are still troubled by considerations that trace to him. This will help clear the way for the exposition of Frege's positive doctrines. As one might expect, they furnish a powerful and attractive philosophy of mathematics, but do not escape unscathed from several technical and philosophical objections.

Frege's Fundamental Distinctions: Sense vs. Reference, Function vs. Object

It is necessary to know something about Frege's semantical doctrines to have a full appreciation of his views on mathematics. As I do not wish to treat his philosophy of language in this work, I will make my account brief. I will present it here so that we may use his semantic terminology freely in the sequel.

Frege's semantics is based upon two interacting ontological hierarchies: the hierarchy of functions and objects, and the hierarchy of senses and references. The former was postulated to explain (among other things) how we can understand sentences we have never heard before, how a group of words can articulate a complete thought, and how compound names manage to refer. Frege introduced the latter to explain why, despite the fact that all identities have the same general truth-condition, certain true identities—such as "the Morning Star equals Venus"—are informative, while others—for instance, "Venus equals Venus"—are trivial. Frege also used the sense-reference hierarchy to rescue the substitutivity of identity and other extensionality principles from apparent failures in so-called intensional contexts. With respect to Frege's philosophy of mathematics, the notions of sense and reference are important for understanding his views on definitions, inference, and analyticity, while his theory of functions and objects is required for a full appreciation of his account of number.

Let us begin the exposition of Frege's theory of sense and reference by considering the following true identities:

(1) The Morning Star equals Venus
(2) Venus equals Venus.

Both are true because the singular terms "Venus" and "the Morning Star," which flank the occurrences of "equals," pick out or refer to the same object. Yet (2) is a triviality and an instance of a law of identity, whereas (1) is both informative and empirical. Indeed, owing to this, someone could believe (2) without believing (1). This, in turn, calls into question the validity of inferences of the form

A believes that Venus equals Venus
The Morning Star equals Venus
Thus, A believes that the Morning Star equals Venus

and casts a shadow on the substitutivity principle for identity.

Frege proposed to handle these difficulties by associating two sorts of meaning with sentences and singular and general terms. A singular term is used to pick out an object; if it succeeds in doing so, then that object is its *reference*. Thus, "Venus," "the Morning Star," and "the Evening Star" all refer to the same planet and have it as their reference. On the other hand, these terms have different senses associated with them—as is evidenced by the fact that people can understand all three terms and

[*19*]

still fail to realize that they refer to the same object. A *sense* is an objective and abstract entity, in Frege's view, which we "grasp" when we understand the word with which it is associated. Each meaningful expression must have a sense, although some such as "Pegasus," "the present King of France," or "$\overset{\infty}{\Sigma}\tfrac{1}{n}$" can fail to have references. An identity is true when and only when its singular terms have the same reference. Thus, (1) and (2) are both true. Yet sentences also have senses and references. The former are the propositions or, as Frege would say, the *thoughts* they express. Frege explained the epistemological differences between (1) and (2) by claiming that they express different thoughts, which he attributed to the differing sense of their component singular terms.

Frege based his theory of the relationship between the meaning of a compound expression and the meanings of its components on the model of mathematical functions. Thus, just as a function always yields the same value when applied to the same argument, a compound expression retains the same reference (or sense) under all replacements of its components which leave their references (or senses) fixed. If we replace "Venus" in (1) by "the Evening Star," we obtain a new sentence,

(3) The Morning Star equals the Evening Star,

which must have the same reference as (1) and (2). But a notable thing that (1), (2), and (3) have in common is that all three are true. Frege thus argued that sentences refer to truth-values, for they remain invariant under all replacements of components by co-referential expressions. Frege also assumed that the thought expressed by a sentence would remain fixed under replacements of its parts by expressions having the same sense.

This pretty theory is threatened, however, by indirect discourse and other intensional contexts. We *can* obtain "John believes that Venus equals the Morning Star" by replacing "Venus" in "John believes that Venus equals Venus" by a co-referential term. But these two sentences may differ in truth-value. Frege responded by arguing that in such contexts the expressions in question have their usual senses as their references. Thus, in the context of our example, "Venus" and "the

Morning Star'' do not refer to the planet, and ''Venus equals Venus'' and ''Venus equals the Morning Star'' *refer* to two different true thoughts rather than to one and the same truth-value. By attributing a systematic equivocacy to ordinary language, Frege parried apparent counterexamples to the substitutivity of identity.

The price is not cheap, however, because the need to handle informative identities of the form ''the sense of *A* equals the sense of *B*'' and iterated intensional contexts appears to commit Frege to a hierarchy of entities to function as higher-level senses. Furthermore, since coextensive but intensionally differing predicates give rise to problems analogous to those created by identities, Frege extended the sense-reference distinction to predicates as well.

(In a letter to Russell, Frege remarked that the ambiguity engendered in English by intensional contexts must be removed in a properly constructed language. This belief would lead to a different analysis of the logical form of intensional contexts. For instance, ''John believes that Venus = the Morning Star'' might be symbolized as

John believes § Venus = § § the Morning Star,

with the result that no term would be used in one context to refer to its ordinary reference and in another to refer to its senses. The resulting formalism would be similar to that of Church's logic of sense and denotation (Church).)

Turning now to Frege's theory of functions and objects, let us begin by noting that Frege's sense-reference distinction did not solve all the problems about language with which he was concerned. He was intrigued also by the question of how we finite beings manage to deal with the infinite variety of expression which our languages make possible. His solution involved the postulation of a structural isomorphism between compound expressions and their senses and references. To each part of a properly analyzed compound expression there corresponds a part in the sense and a part in the reference of the compound expression. These ''parts'' are the senses and references of the component expressions in question. For example, since the sentence

2 is a prime

would be analyzed into "2" and "is a prime," the thought it expresses and the truth-value to which it refers would be analyzed into parts that correspond to "2" and "is a prime." But for this approach to provide an explanation of the infinite variety of language, it must be possible for these parts to be recombined to form other compounds. Frege was thus faced with the question of explaining how these separable parts combine into unified wholes.

In proposing his solution, Frege played heavily upon the fact that sentences, proper names, and definite descriptions can be used by themselves to make independent linguistic acts (to answer questions, for example) while predicative or functional expressions, such as "is red," "is greater than 2," "plus one," usually cannot be so used without further linguistic supplementation. (Frege would claim that they never can be so used, but his grammatical classification may beg the question.) Frege then argued that the senses and references of these incomplete expressions also must be incomplete or unsaturated. He believed that this incompleteness provided them with the ability to combine with other senses and references into united wholes. I do not wish to minimize the importance of Frege's linguistic insights, but his investiture of certain senses and references with unifying powers hardly constituted a genuine solution to the problem that motivated his theory.

Frege further held that there are differing types of incompleteness or unsaturatedness so that an unsaturated expression, sense, or reference can be completed only by an expression, sense, or reference of the appropriate kind. Proper names, sentences, and complex numerical terms all count as complete expressions. Their references are called *objects* and they are generally called object-names. If an object-name is deleted from an object-name in which it occurs, then we obtain a *first-level function-name* of degree one. If two different object-names are deleted, then we obtain a first-level function-name of degree two; and so on. For example, starting with the object-names

$$2 + 3, \quad 5 > 5,$$

we may obtain first-level *function-names,* which we write as

$$\xi + 3, \quad \xi > \xi, \quad \xi + \zeta,$$

[22]

where the Greek letters mark the gaps or incomplete places in these names. These refer to first-level *functions*. The reference of a complex object-name (consisting of a function-name and object-names that complete it) is the value of the function to which the function-name refers for the references of the object-names as arguments. The sense of a compound expression is similarly a function of its parts. First-level functions of degree one whose values are truth-values, such as the function $\xi > 7$, are called *concepts;* and an object *falls under* a concept if, and only if, its value for that object as argument is true. Frege associated an *object* with each first-level function of degree one called its *value-range*. The value-ranges of two functions are identical when and only when the functions have the same values for every argument. The value-range of a concept is its *extension*, and it plays the role of a set or class, since two concepts will have the same extension when, and only when, the same objects fall under both.

Frege extended his hierarchies of expressions, senses, and references to higher levels by analyzing quantifiers as higher-order predicates. Consider the sentence

$$(\exists x)\,(x > 5).$$

We can delete "5" from it and obtain a first-level function-name. But what happens when we delete the first-level function-name "$\xi > 5$" from it? If we write the result as

$$(\exists x)\phi x,$$

we have obtained an incomplete name that can be completed not by object-names but only by first-level function-names (of degree one). Thus, we must recognize *second-level functions* as the references for such names and higher-level intensions as their senses. By extending this analysis to quantification over first-level functions, Frege arrived, in turn, at *third-level functions*. Evidently, the hierarchy can be extended indefinitely.

Frege's thesis that compound referring expressions are possible only if their incomplete parts designate (incomplete) functions has three important consequences for his views in logic and the philosophy of mathematics. *First,* predicates must be viewed as designative, and thus quantified predicate variables are a legitimate part of logic. *Second,* the hierarchy of incomplete entities permeates all language; no proper language can violate it. In particular, no variable may range over both functions and objects; so standard methods for "reducing" many-sorted logics to one-sorted ones cannot be legitimately applied to Frege's theory. These two consequences commit Frege inextricably to second-order logic (and possibly to higher-order logics as well). *Third,* the extension of a concept and the concept itself must be sharply distinguished, since the former is complete whereas the latter is not. Thus, Frege found himself in need of both second-order logic and set theory. This overly rich system quickly gave rise to the Russell paradox, which eventually discredited Frege's system.

I shall return to the matter of Frege's conception of set theory and logic when I examine Frege's views in Chapter Five. The reader who is interested in learning more about Frege's hierarchies of incomplete entities can consult Dummett (1973), Klemke, and Thiel. The discussion I have just presented is elaborated and documented in Resnik (1965).

ONE

Psychologism

From the prevailing logic I cannot hope for approval, . . . for it seems to be thoroughly infected by psychology. If people consider, instead of things themselves, only subjective representations of them, only their own mental images—then all the most delicate distinctions in the things themselves are naturally lost and others appear instead which are logically quite worthless.—Frege 1893:xiv

And that is how our thick books of logic came to be; they are puffed out with unhealthy psychological fat which conceals all finer forms. Thus a fruitful collaboration of mathematicians and logicians is made impossible. While the mathematician defines objects, concepts and relations, the psychological logician watches the becoming and changing of ideas, and at bottom the mathematician's way of defining must appear to him just silly, because it does not reproduce the essence of ideation. He looks into his psychological peepshow and says to the mathematician: "I cannot see anything at all of what you are defining." And the mathematician can only reply: "No wonder, for it is not where you are looking for it."—Frege 1893:xxiv

These sarcastic passages are typical expressions of Frege's disdain for the "psychological" approach to logic and the philosophy of mathematics which dominated the thinking of his philosophical contemporaries. Their analyses were valueless as foundations for logic and mathematics, as Frege, armed with a logic extending far beyond the comprehension of most of his philosophical colleagues, was acutely aware. Yet their reputation so tarnished Frege's work that his fellow mathematicians also ignored it, and this in turn produced in Frege the deep sense of rejection and isolation that comes through in these and the surrounding passages.

Unfortunately, this clouded Frege's critical faculties—not to the extent that his arguments missed the mark or that his philosophical colleagues produced insights that he failed to see, but rather to the extent that he tended to conflate a great variety of views under the heading of subjectivism or psychologism. I shall not attempt to disentangle all of them here or to trace their roots in British empiricism or in German idealism and historicism. Instead, I shall make a step toward clarifying the situation by considering four positions that Frege criticized, presenting in each case an exposition of the views of a representative of the position and a discussion of Frege's critique of it. The four themes to be considered are (1) the use of mental entities in place of abstract ones, (2) the preference for descriptions of the genesis of mathematical and logical notions over reductive definitions, (3) the treatment of logic as the science of human thinking, and (4) the reduction of truth to acceptance. It should be noted that although most views that contain one of these themes contain several others, the themes themselves are independent. I shall leave to others the task of explaining why and how these themes came to pervade the work of the philosopher-logicians of Frege's time. (To my knowledge, this task has been neglected by current histories of logic, which focus instead on the mathematician-logicians of that period. In Sluga (1976) there is some discussion of the intellectual background to psychologism.)

The Substitution of Mental Entities for Abstract Ones

By an abstract entity I shall mean an entity that is neither spatial nor temporal and is neither a product of nor a process in a mind. Many contemporary philosophers follow Frege and view classes, numbers, and meanings as typical abstract entities, but Frege's contemporaries were loath to recognize them. Whereas Frege thought that coherent accounts of mathematical truth and the semantics of natural and formal languages would be impossible without their introduction, his opponents attempted to develop views in which physical or mental phenomena performed the functions for which Frege introduced abstracta. Although in the end I shall agree with Frege that these attempts are quite faulty, I cannot deny that the epistemology for abstract entities was and is still pretty much a mystery. For it is very difficult to understand how we can come to have knowledge about things that exist without us and yet are incapable of (physically) interacting with us. Thus, there is considerable motivation for developing views that dispense with such entities.

The particular theses with which we shall be concerned here are the claims (a) that numbers and other mathematical entities are ideas and (b) that senses or intensional meanings are ideas. I shall follow Frege's use of the word "idea," taking it to be a subjective mental content, and assume with him—at least for the purpose of the present discussion—that we cannot have direct access to anyone else's ideas and, consequently, that ideas must be taken as ideas of a particular individual, thinking being and of no one else (Frege 1884:sec. 26; 1893:I, p. XVIII; 1892a:29).

Although Frege accuses a number of writers of holding some variant of theses (a) and (b) and criticizes the views of Benno Erdmann (Frege 1893:I, pp. XV-XXV) and Edmund Husserl (Frege 1894) at length, I have chosen the writings of John Locke as the source for my exposition of these two claims. Locke's works are more accessible to English-speaking readers—neither Erdmann's *Logik* nor Husserl's *Philosophie der Arithmetik* have been translated, and copies of the former are difficult to obtain—and although it may be debatable whether anyone ever held exactly the views that Frege attacked, the passages from Locke which I shall cite certainly appear close to them. (Ironically, Frege quotes Locke in the *Grundlagen* but does not name him when he attacks subjective views.)

In his *An Essay Concerning Human Understanding*, Locke took his main goal to be that of showing that all knowledge derives from experience:

> Our observation, employed either about external sensible objects, or about the internal operations of our minds, perceived and reflected on by ourselves, is that which supplies our understandings with all the *materials* of thinking. These two are the fountains of knowledge, from whence all the ideas we have, or can naturally have, do spring. [Locke:Bk. II, chap. I, sec. 2]

He presented long and vigorous arguments against the possibility of innate knowledge (Locke:Bk. I) and never entertained the possibility of nonsensuous perception of external objects. This left him with the two sources of concepts and knowledge mentioned in the passage— sensation and reflection, where the latter produces new ideas through the observation of our own thought processes (Locke:Bk. II, chap. I, secs. 3–5). Abstract entities were thus effectively excluded from Locke's philosophical system at the outset.

Locke uses the term "idea" as Frege did, to signify a mental entity:

> I have used [the term "idea"] to express whatever is meant by *phantasm, notion, species, or whatever it is which the mind can be employed about in thinking....*
>
> I presume it will be easily granted me, that there are such *ideas* in men's minds. [Locke: Introduction, sec. 8]

However, Locke's theory would not be subjectivistic if different men could have identical ideas. Yet, like Frege, Locke assumes that this is something which never could be known "because one man's mind could not pass into another man's body, to perceive what appearances were produced by [his] organs" (Locke:Bk. II, chap. XXXII, sec. 15). Unlike Frege, Locke maintained that at least "the sensible ideas produced by any object in different men's minds, are most commonly very near and undiscernibly alike" (ibid.).

Locke's discussion of numbers is directed chiefly to showing how the positive whole numbers may be constructed from our simple (or unanalyzable) idea of *unity:*

> Amongst all the ideas we have, as there is none suggested to the mind by more ways, so there is none more simple, than that of

unity or one. It has no shadow of variety or composition in it: every object our senses are employed about, every idea in our understandings, every thought of our minds, brings this idea along with it. [Locke:Bk. II, chap. XVI, sec. 1]

The other numbers are formed by repeating this idea to form complex ideas which then are designated by number words:

> Thus, by adding one to one, we have the complex idea of a couple; by putting twelve units together, we have the complex idea of a dozen; and so of a score, or a million or any other number. [Locke:Bk. II, chap. XVI, sec. 2]

> By the repeating, as has been said, the idea of an unit, and joining it to another unit, we make thereof one collective idea, marked by the name two. [Locke:Bk. II, chap. XVI, sec. 5]

Unfortunately, Locke confuses a unit with the number one, a dozen with the number twelve, and so on. While the expressions "twelve" and "a dozen" are interchangeable in many contexts (e.g., "a dozen eggs," "twelve eggs") other contexts indicate that, on the face of it, "twelve" designates a definite object of mathematical study (e.g., "twelve has two prime factors") while "a dozen" characterizes classes of twelve things. Locke's theory also suffers from his failure to distinguish between a number and our idea of it. For even if the number two is itself an idea, we can associate very different ideas with it. (These criticisms are Fregean. See Frege 1884:secs. 37, 38; 1893:I, pp. XVIII, XXII.) These unclarities in Locke's views make it difficult to determine whether his numbers are obtained by abstracting them from collections or whether they are obtained by mentally combining previously generated numbers. Nonetheless, the theme that numbers are ideas produced by means of our own mental operations comes through quite clearly. It is to this that Frege's chief criticisms are directed.

Let us turn now to Locke's theory of meaning. Language, he tells us, is used primarily to communicate our ideas to others. To this end we use words to signify ideas: "Words in their primary and immediate signification, stand for nothing but *the ideas in the mind of him that uses them*" (Locke:Bk. III, chap. II, sec. 2). We do assume that other people also use words to signify their own ideas and that words can stand also (in a secondary sense) for "the reality of things" as well as ideas. Through the repeated use of a given word to signify a certain idea, an association between the word and the idea is created so that when someone else utters the word, the idea we associate with it will be called to mind. Thus, people communicate by uttering the words which they associate with the ideas they wish to make known. The words uttered will then evoke the associated ideas in their audience. For communication to work, of course, the speaker and his audience must associate sufficiently similar ideas with the words in question. This will happen when they speak the same language (Locke:Bk. III, chap. II, secs. 4–7).

It is clear that this theory has to cope with many problems—and Locke *did* attempt to solve some of them. Since they form the basis for Frege's criticisms, we will see Locke's theory in a clearer light if we present its further developments as responses to Frege.

An ideational theory of meaning is unworkable, Frege argued, because it conflates the truly subjective aspects of meaning with its public ones. For example, utterance of the sentence "Pegasus does not exist" is likely to evoke images in a horseman quite different from those it would evoke in a specialist in ontology. Yet both the ontologist and the horseman can agree on what has been said and its truth (Frege 1892a:29). Furthermore, without a distinction between subjective and objective meaning, logic would be paralyzed; for in order to evaluate inferences, it must be possible to paraphrase sentences while preserving their logically essential features, even if this alters their more subjective features, such as poetic imagery (Frege 1879b:sec. 3; 1892b:196 note; 1918:63–64). Finally, there are many words, such as prepositions and sentential connectives, with which we associate no ideas whatsoever. Yet we cannot deny their meaningfulness as Locke would have us do (Frege 1884:sec. 60).

Locke did not deal with those specific criticisms, but there is no problem in seeing how he could have replied to them. First, he might have denied that *every* word is meaningful when considered in isolation, attributing meaning only to certain parts of speech—for example, to singular and general terms and sentences. Frege hardly could have objected to this reply, since his theory of meaning is based upon a similar assumption. Indeed, in the *Grundlagen* he took sentences as the basic unit of meaning and accused psychologism of being driven to the ideational theory by its quest for meanings for words (Frege 1884:sec. 60, p. x). This position was modified in his later works, with the result that singular and general terms were also attributed meaning. Yet Frege's treatment of variables makes it clear that even in his later works he would not attribute a meaning to every word or symbol (Frege 1904:659).

Locke could have replied to the Fregean objection concerning the confusion between objective and subjective meaning by applying his treatment of general terms and abstract ideas to the problem of meaning (Locke:Bk. III, chap. III). For just as we can abstract from our ideas of particular horses to form the general idea *horse,* so can we abstract from the particular ideas occasioned by the utterances of a given word to obtain a more general idea which can function as its meaning. Indeed, Locke's discussion of how we learn general terms fits this picture exactly (Locke:Bk. III, chap. III, secs. 6–7).

Frege's response to this might have taken the form of a parody of abstraction which reveals the difficulties that arise when one confuses an account of how we learn to use general terms with an account of the relationship between a general term and the things to which it applies (Frege 1894:316–317). However, although Frege is correct in his claim that we do not form, say, the concept *man* by changing individual men, Locke distinguished between individual men and our ideas of them, and it is only upon the latter that we operate when we form general ideas. Thus, the bearing of Frege's critique of abstraction upon Locke's view is somewhat tangential.

Far more to the point is Frege's often repeated objection that the privacy of meanings under the ideational theory makes genuine communication and disputes over the truth of statements impossible:

> If the same thought is not taken by me and by [another] as the content of the Pythagorean Theorem, then properly one should not say "the Pythagorean Theorem" but rather "my Pythago-

rean Theorem'', ''his Pythagorean Theorem'', and these would be different. . . .

If every thought requires a bearer to whose consciousness it belongs, then it is a thought of only its bearer. There would be no science common to many on which many could work, but perhaps I would have my science, namely, a totality of thoughts of which I would be the bearer, another would have his science. Each of us would deal with the content of his own consciousness. A contradiction between the two sciences would not be possible, and it would be essentially pointless for us to dispute over truth. [Frege 1918:68-69; see also Frege 1884:sec. 27; 1892a:29-30; 1893:I, p. XVIII; 1894:317]

Locke's *Essay* contains several sections that attempt to deal with this and similar objections. The most extensive of these is chapter four of Book Four, entitled ''Of the Reality of Knowledge.'' He argues there that our simple ideas must conform to reality, since they are not of our own making but rather are products of our interactions with external things. Furthermore, our complex ideas—with the exception of those of substances—are not intended as representatives of an external reality, ''so that we cannot but be infallibly certain, that all the knowledge we attain concerning these ideas is real, and reaches things themselves [namely, the ideas]'' (Locke:Bk. IV, chap. IV, sec. 5). In particular, mathematical knowledge is for this reason both real and certain:

I doubt not but it will be easily granted that the knowledge we have of mathematical truths, is not only certain but real knowledge; and not the bare empty vision of vain, insignificant chimeras of the brain: and yet, if we will consider, we shall find that it is only of our own ideas. [Locke:Bk. IV, chap. IV, sec. 6]

Even if we grant Locke everything he has claimed, he still falls short of answering Frege's objection—although the opening passages of the chapter, which discuss the contradictory claims of a wise man and a dreamer, demonstrate that he has this sort of objection in mind. For Locke's rebuttal at most establishes that each of us is capable of certain and genuine knowledge of specific kinds, and it fails to establish that this is a common body of knowledge.

If each of us has a different store of ideas, then the Fregean objection appears right on the mark. For even if I am certain that my ideas verify what I express with the Pythagorean Theorem, I have no basis for claims about what your ideas do. For that matter, what if each of us reacted differently to the external world, with the result that each of us had differing simple ideas? Would it then not follow that none of our ideas would be comparable, since all our complex ideas are constructed from simple ones? Here is Locke's retort:

> Neither would it carry any imputation of falsehood to our simple ideas, if by the different structure of our organs it were so ordered that *the same object should produce in several men's minds different ideas* at the same time; e.g. if the idea that a violet produced in one man's mind by his eyes were the same that a marigold produced in another man's, and *vice versa*. For, since this could never be known, because one man's mind could not pass into another man's body, to perceive what appearances were produced by those organs; neither the ideas hereby, nor the names, would be at all confounded, or any falsehood be in either. For all things that had the texture of a violet, producing constantly the idea he called blue, and those which had the texture of a marigold, producing constantly the idea which he as constantly called yellow, whatever those appearances were in his mind; he would be able as regularly to distinguish things for his use by those appearances and understand and signify those distinctions marked by the name blue and yellow, as if the appearances or ideas in his mind received from these two flowers were exactly the same with the ideas in other men's minds. [Locke:Bk. II, chap. XXXII, sec. 15]

It is interesting to compare this with a similar passage from Frege:

> Space, according to Kant, belongs to appearance. For other rational beings it might take some form quite different from that

in which we know it. Indeed, we cannot even know whether it appears the same to one man as to another; for we cannot, in order to compare them, lay one man's intuition of space beside another's. Nevertheless, there is something objective in space all the same; everyone recognizes the same geometrical axioms, even if only by his behavior, and must do so if he is to find his way about the world. What is objective in it is what is subject to laws, what can be conceived and judged, what is expressible in words. What is purely intuitable is not communicable. To make this clear, let us suppose two rational beings such that projective properties and relations are all they can intuite—the lying of three points on a line, of four points on a plane, and so on; and let what the one intuites as a plane appear to the other as a point, and vice versa, so that what for the one is the line joining two points for the other is the line of intersection of two planes, and so on with the one intuition always dual to the other. In these circumstances they could understand one another quite well and would never realize any difference between their intuitions, since in projective geometry every proposition has its dual counterpart; any disagreements over points of aesthetic appreciation would not be conclusive evidence. Over all geometrical theorems they would be in complete agreement, only interpreting the words differently in terms of their respective intuitions. With the word "point," for example, one would connect one intuition and the other another. We can therefore still say that this word has for them an objective meaning, provided only by this we do not understand any pecularities of their respective intuitions. [Frege 1884:sec. 26]

We can generalize the idea contained in these two passages by thinking of two people who use a first-order language with the same syntax but interpret their language via differing isomorphic interpretations, so that a predicate under the one person's interpretation is true of an ordered n-tuple of objects just in case the same predicate under the other person's interpretation is true of the isomorphic image of this n-tuple. It would then follow that each sentence would be true under one interpretation just in case it was true under the other. (Cf. Mendelson:90–91.) Then if both people in question had genuine knowledge concerning the truth-value of a sentence—even if it is private knowledge—they could not but agree concerning its truth or falsity.

One could supplement this linguistic account with a psychological claim to the effect that people are so constituted that their private ideas or intuitions will be isomorphic. Frege's reference to Kant and his restriction of the discussion to space undoubtedly means that he presupposed that an isomorphism concerning spatial intuitions would be guaranteed by the Kantian forms of intuition. Locke appears to assume that we all have the same set of simple ideas due to our similar biological makeup, but in the passage in question he seems to take back that assumption. In any case, there is no argument to be found in Locke for the conclusion that given that we have the same set of simple ideas, they will be so correlated with words to produce isomorphic models.

In view of Frege's tirades against subjective theories of meaning, one cannot help wondering what "objective meanings" he is referring to in the geometrical example. A natural move would be to identify the objective reference of a term with the class of all its isomorphic subjective interpretations. Thus, the term "point" of the example would refer to the class {$point_1$, $plane_2$} and the term "plane" would refer to {$plane_1$, $point_2$}. The subscripts, which indicate whose points and planes are involved, would prevent the two terms from being assigned the same reference despite their duality. Perhaps the objective sense of the terms could be defined in terms of their role in inferences. This would fit in well with Frege's definition of number in terms of equivalence classes and with some of his remarks on sentences which from a logical point of view express the same thought.

As a general approach to the theory of meaning, however, the isomorphism view is not satisfactory. There appears to be much evidence against it in the form of speakers of the "same language" who both claim to be fully informed and yet disagree about the truth of a given sentence. For what could be better evidence of nonisomorphic differences of meaning than irresolvable disagreements over truth? One cannot save the situation by restricting agreement to a set of core sentences (meaning postulates). This would mean simply abandoning the isomorphism theory without providing for "objective" interpretations of the non-core sentences.

Turning now to Locke's theory of number, we see that the Fregean no-common-science objection applies at the level of reference as well as sense:

> It is impossible to ascribe to every person his own number one; for in that case we should first have to investigate the extent to which the properties of these ones agreed, and if one person said "one times one is one" and the next said "one times one is two", we could only register the difference and say: your one has one property, mine has another. There could be no question of any argument as to who was right, or of any attempt to correct anyone; for they would not be speaking of the same object. [Frege 1893:I, p. XVIII]

Naturally, any attempt by Locke to reply that our ideational number series are isomorphic because they are constructed similarly from our ideas of unity would run into the difficulties already enumerated. In addition, the fact that the method of verification is introspective and cannot be checked by others seems to make the fact of our agreement a matter of coincidence unless, of course, we explain it in terms of a common psychology or conditioning. In the latter case we seem forced to classify someone who disagrees with us as psychologically or socially aberrant rather than as simply wrong. Fortunately, Frege seems to have restricted this approach to geometry. For he showed no such tolerance in the case of logic:

> But what if beings were even found whose laws of thought flatly contradicted ours and therefore frequently led to contrary results even in practice? The psychological logician could only acknowledge the fact and say simply: those laws hold for them, these laws hold for us. I should say: we have a hitherto unknown type of madness. [Frege 1893:I, p. XVI]

It might be objected that my counterexample is vacuous, since no two people can be *fully informed* about the facts of a matter and yet disagree. Granted, if the facts are objective. But in the case in point, each party is fully informed about his own ideas and knows nothing of the other's. And here, as Frege said, disputes over the truth are idle.

Another serious Fregean objection to the ideational theory of number is that it makes arithmetic dependent upon human existence and psychology:

> Astronomers would hesitate to draw any conclusions about the distant past, for fear of being charged with anachronism,—with reckoning twice two as four regardless of the fact that our idea of number is a product of evolution and has a history behind it.
> [Frege 1884:VI]

In other words, $2 \times 2 = 4$ in every possible world—even those in which human beings do not exist. This is an extremely compelling argument for *someone who already leans toward an objective view of number theory* since it contains one of the key insights of the objective theories. But someone who regards numbers as human creations need not accept its full force. For in a possible world without numbers—if there be such—the facts inferred from applied number theory would still obtain. For instance, it would still be the case that if I have one penny in my left pocket and one dime in my right pocket, then I have at least two coins in my pockets. The reason for this is that such propositions can be reformulated by replacing the use of numerical terms by numerical quantifiers. The example given can thus be recast as:

> If there is one and only one penny in my left pocket
> and there is one and only one dime in my right pocket
> then there are distinct coins x and y in my pockets
> and no others.

Furthermore, the numerical quantifiers themselves can be defined in first-order logic with identity—something first noted by Frege (1884:sec. 55).

This does not take care of possible worlds in which beings create numbers that violate the laws of arithmetic, but advocates of an ideational theory simply should deny this possibility. It is essential to their view of the number series that it be generated inductively—that is, each numerical idea is obtained from its predecessor by applying the mental operation of "adding one." But, as is well known, every inductively generated series satisfies the axioms of number theory. Frege's opponents thus could argue to a standoff concerning the dependence of numbers on human psychology.

A third objection of Frege's requires a more elaborate response. The objection is that the ideational number series terminates with the last number that someone has constructed, whereas mathematics requires an infinite number series (Frege 1884:sec. 27; 1894:328–329). Husserl dealt with this problem in some detail, for he believed that human beings have no proper ideas of numbers beyond the number twelve (Husserl:192). His theory of symbolic ideas attempts to show how the notation of number theory can extend our ideas of number beyond those we can form directly. Unfortunately, Husserl never adequately explained what a symbolic number is and how it differs from a real one, and his theory appears to be another variant of formalism (Frege 1894:328–331). Locke never faced the Fregean objection directly but argued that we do have a clear idea of the number series as an unending one, since reflection informs us that for every number we have formed, a successor can be formed by adding one. He denied, however, the possibility of obtaining a coherent idea of either infinite numbers or a completed number series (Locke:Bk. ii, chap. xvii). In short, Locke recognized the potential infinite while disavowing infinite totalities. Locke was, of course, unaware that this would require changing the face of mathematics considerably. We were taught this only during this century by the critique of the use of infinite totalities in mathematics furnished by contemporary constructivists (Bishop; Heyting). An adequate response to the third objection would require the adoption of constructivism.

By way of reviewing our discussion of Frege's critique and the Lockean responses, let me cast the ideational theory in the form of an apology for constructive mathematics. This will serve not only to focus our attention on issues that are still of interest but also to further dramatize the conflict between "subjectivists" and "objectivists." The apology begins with the observation that classical (i.e., nonconstructive) mathematical reasoning is founded upon a conception of mathematics as dealing with a preexisting infinite domain of objects whose properties are independent of human thought and activity. Thus, to appeal to classical mathematics to underwrite a Fregean conception of mathematical objects is to beg the question. The same question is begged by appealing to the eternality of mathematical truth, since in constructive mathematics new truths are continually being created by human mathematical activity. Finally, without assuming common mathematical objects, constructivism can be based upon intersubjective mathematical activity with an intersubjective conception of correctness. For we are each so constituted that we can be conditioned by mathematical training to form our own inductive sequences of mental constructions. These, our natural numbers, by being isomorphic models of number theory, will form the basis for our intersubjective activity.

I do not wish to endorse this apology—no constructivist has succeeded in giving us a plausible account of the psychology assumed in it. Nor do I wish to imply that Frege's arguments are without force. I only wish to point out that considered in isolation they beg the question somewhat. In the context of a full theory of meaning and truth, coupled with a plausible epistemology for abstract entities, they could be quite compelling. Unfortunately, as will be seen in a later chapter, Frege never developed the epistemology requisite to his views.

The Preference for Descriptions of the Genesis of Mathematical and Logical Notions over Reductive Definitions

I shall take Husserl's *Philosophie der Arithmetik* as my source for this theme in psychologism. Although we have already seen the psychogenic point of view at work in Locke's attempt to trace all our ideas to sense experience, I think it will be useful to discuss the views of someone who was directly engaged in philosophical argumentation with Frege. Not only did Frege and Husserl criticize each other, but much of their exchange centered around Frege's definition of number. From this we shall be able to mine issues that remain very much alive today.

Husserl's book has many similarities to Frege's *Grundlagen*. Both books take the analysis of the concept of number to be a fundamental task for the philosophy of mathematics and general epistemology. Both contain extensive criticisms of the views of their predecessors and contemporaries, and both attempt their own positive accounts. Furthermore, both display the same style of argumentation.

On the other hand, Husserl's work is a psychological as well as a logical investigation and, as a result, contains many elements and conclusions that were inimical to Frege's way of thinking. I shall not present Husserl's positive analysis here. Some of it will be evident in the passages I shall quote, the rest seems to me accurately summarized and criticized in Frege's review (Frege 1894). In any case, the details of his analysis are not germane to our discussion.

It is only natural to inquire why Husserl should arrive at a point of view so different from Frege's when the books have so much in common. The reason is that Husserl begins with the premise that the concept of number is incapable of definition and that, therefore, the only appropriate way to clarify it is to trace its psychological origins. The following passage, which was written with explicit reference to Frege and his program, illustrates this view:

> One can only define the logically complex. As soon as we come upon ultimate elementary concepts all defining ends. No one can define concepts such as quality, intensity, place, time, and the like. And the same holds for the elementary relations grounded on these concepts. Identity, similarity, increase, whole and part, plurality and unity, and so on, are concepts which are entirely unfit for formal-logical definitions. What we can do in such cases consists only in indicating the concrete phenomena from which they are abstracted and clearly laying out the type of abstraction processes used. . . . From a rational point of view we can demand that the linguistic presentation of such concepts be fixed accordingly: it must be well arranged for us to place ourselves in the right disposition to display to ourselves the abstract moments in inner or outer intuition which are meant [by them] or to recreate for ourselves the psychological processes which are required for the formation of these con-

cepts. This will only be useful and necessary if the name signifying a concept does not suffice for understanding it—be it due to extant equivocations, be it due to any kind of misinterpretation. The concept of number is exactly such a case and we cannot find them blameworthy when mathematicians, at the initial point of their system, instead of giving a logical definition of the concept of number, "describe the way in which we arrive at this concept''; only the description must be accurate so that it will also fulfill its purpose.

Our previous analyses have established with uncontradictable clarity that the concept of plurality and unity rest upon ultimate elementary psychological data and thus must belong with concepts which are undefinable in the indicated sense. But the concept of number is so closely connected with these that one can scarcely speak of defining it either. The goal which Frege set himself is to be called a chimerical one. It is thus no wonder that his work, despite all its acumen, lapses into unfruitful hypersubtleties and ended without positive results. [Husserl:119–120; see also p. 96]

On the other hand, Husserl took the extension of the concept of number to be unproblematic, since no one has any difficulty in applying it to given cases. The analysis of the concept of number thus must be directed at its content (Husserl:15). But Frege's definition only characterizes that extension of the concept of number and thus falls short of the goal of a philosophical analysis. This is why Husserl said the following regarding Frege's definition and his method of converting equivalence relations into equivalence classes:

I cannot see that these methods signify an enrichment of logic. Their results are of such a type that we can only wonder how anyone could take them as true even if only temporarily. In fact what these methods permit us to define are not the contents of the concepts of direction, form, number, but rather their extensions. . . . all these definitions would become true propositions if for the concepts to be defined their extensions were substituted, but, of course, they would become entirely self-evident and worthless propositions. [Husserl:122]

Why do Frege's definitions fail to characterize the content of the concept of number as well as its extension? Here are Husserl's grounds:

> If numbers are defined by means of the relation of equivalence [i.e. one-one correlation], then every assertion of number is always directed—instead of at the concrete set at hand—at the relationship of the same to other sets. To attribute a definite number to this set means classifying it with a definite group of equivalent sets; but this is not at all the sense of an assertion of number. Let us consider an example. Do we call a set of nuts lying before us four, because it belongs to a certain class of infinitely many sets which can be put into mutual unique correspondence? Of course, no one ever had such thoughts when doing that, and we can scarcely find any practical motives at all for us to be interested in the same. What in truth interests us is the circumstance that there is one nut and one nut and one nut and one nut. [Husserl:116]

Husserl also rejected the characterization of numerical equality in terms of one-to-one correlation on similar grounds; for while the possibility of correlating the F's with the G's is a necessary and sufficient condition for the number of F's to equal the number of G's, "the knowledge that the two numbers are equal does not at all require the knowledge of the possibility of the correlation" (Husserl:105).

To summarize, Husserl took number to be a logically simple and thus undefinable concept. As a consequence, he maintained that the only analysis of which it is susceptible is a psychological analysis of the process through which we arrive at this concept. A successful account of this not only would remove the philosophical difficulties associated with number but would provide us also with a sufficiently clear concept for use in science and mathematics. Frege's attempts failed in this regard because they furnished only extensional characterizations—which were unnecessary in the first place.

Frege's long review of Husserl's book attacks a number of points in addition to the ones just summarized. He criticized Husserl for conflating the objective and subjective senses of the word "representation" (or "idea": the German is *Vorstellung*) (Frege 1884:318). Then, after he saddled Husserl with a subjective ideational theory of numbers, Frege brought his usual criticisms of such theories to bear upon it (Frege 1894:315–321). He also attacked the notion of mental abstraction (Frege 1894:323–325); Husserl's analysis of number (Frege 1894:321–322); and his treatment of zero, one, and the infinity of the number series (Frege 1894:324–332). However, only a few paragraphs are devoted to Husserl's criticisms of Frege's own program. Part of the reason, to be sure, is that Frege took the criticisms to be consequences of Husserl's (alleged) ideational theory:

> If words and combinations of words denote ideas, then with two of them nothing further is possible except that they denote the same idea or different ones. In the first case it is pointless to postulate their identity through a definition; "an obvious circle"; in the other case it is false. This is the sort of objection which the author regularly advances. A definition may not analyze the sense [of an expression] either, since the analyzed sense is not the original one. Either I already think everything clearly with the expression to be defined, then we have an "obvious circle"; or the defining expression has a richer analyzed sense, then I do not think with it the same thing as I do with the one to be explained: the definition is false. . . . Obviously there is a parting of the ways here between the psychological logicians and mathematicians. The former take the senses of words and ideas as the main thing, the others, on the contrary, the things themselves, the references of the words. [Frege 1894:319–320]

Yet, after giving us such a nice statement of the paradox of analysis and dismissing it as a consequence of the ideational theory, Frege continued with words which raise even deeper questions:

> The objection that it is not the concept but its extension that is defined really touches all mathematical definitions. For the

mathematician, it is no more right and no more wrong to define a conic section as the line of intersection of a plane with the surface of a circular cone than to define it as a plane curve with an equation of the second degree in Cartesian coordinates. Which definition he chooses—one of these two, or some other again—depends entirely on reasons of convenience; although the expressions neither have the same sense, nor evoke the same images. I do not mean by this that a concept and its extension are one and the same; but coincidence in extension is a necessary and sufficient criterion for the occurrence between concepts of the relation corresponding to identity between objects. [Frege 1894:320]

Husserl's objections raise a number of serious issues for Frege's program. Most of these issues received adequate answers in other portions of the Fregean corpus, and I will postpone a detailed exposition of them until I present Frege's positive views in a later chapter. It is useful, however, to indicate the issues here, since Frege's review tends to overlook them in its zeal to demolish Husserl's psychologistic analyses. A more sympathetic reading of Husserl would have made Frege deal more fully with the following questions: (1) the purpose of (mathematical) definitions, (2) the explanation of primitive terms, (3) the clarification of concepts by means other than definitions, and (4) the nature of concepts. All of these are fundamental concerns of logic or the philosophy of mathematics which arise independently of subjective theories of abstract entities. I shall mention Frege's views on each briefly now.

Let us start with (4), since the last Fregean passage quoted does hint at Frege's treatment. As was explained in the Introduction, Frege viewed concepts as abstract, mind—and language—independent entities to which we refer by means of predicates. Since they can be the references of predicates, they must be subject qua references to the same sort of extensionality principle as objects. So, just as we can interchange proper names (in extensional contexts) while preserving the reference of the total context in which they occur, if and only if they are coreferential, so also must we be able to exchange predicates in an extensional context in which they occur just in case they are co-referential.

This, together with Frege's identification of the references of sentences with truth-values, implies that co-extensive predicates are co-referential. Otherwise, truth would not be preserved in passing from "$Fa \equiv Fa$" to "$Fa \equiv Ga$", where F and G are coextensive. The additional step of postulating *a unique reference* to which co-referential predicates refer and identifying this with a concept guarantees that coincidence in extension is an "identity" condition for concepts. However, Frege's hierarchy of functions will not allow him to identify the identity relation that holds among objects with the "identity" relation that holds among concepts. Since even the expression "$\xi = \zeta$" cannot be completed by predicates, another expression of the appropriate type must be introduced. In discussing this in a manuscript written between 1892 and 1895, Frege introduced the expression

$$\phi(\alpha) \underset{\approx}{\overset{\alpha}{}} \psi(\alpha)$$

for this purpose. Here the letter α serves to fill the argument places of the incomplete expressions substituted for ϕ and ψ (Frege 1969:131–132).

To pass on to topic (1), we can now see why Frege argued that characterization of the extension of a concept was sufficient for picking out the concept itself. But surely Husserl could reply that regardless of what we call it, a proper analysis of a predicate should pick out its intensional meaning or, in Frege's terms, its sense. Frege's answer to this could take its start from the words "*For the mathematician* it is no more right and no more wrong to define a conic section as . . ." of the passage given above. I have emphasized the phrase "for the mathematician" because Frege must fall back upon the extensionality of the language of mathematics. Since the definitions in question have coextensive *definiens,* every statement about conic sections will hold under one definition which holds under the other. Mathematics will be unable to note a difference between the definitions, so that neither will be "more right or more wrong."

Frege did not take into account the existence in mathematics of definitions that are "no more right or no more wrong" and yet are not even extensionally equivalent. The real numbers, for example, have been defined as Dedekind cuts, the upper or lower members of such cuts, and equivalence classes of convergent sequences or rational numbers, to name but a few of their definitions. Yet, no real number defined as a Dedekind cut, for instance, is identical with a real number construed as a lower member of a cut, since the former is an ordered pair while the latter is a member of such pairs. Other examples abound in mathematics, but in Chapter Five we will be interested particularly in the bearing of alternative definitions of numbers as sets upon Frege's problem.

For the present, let us take note of two points. First, Frege's response to Husserl was inadequate even with respect to the definition of mathematical concepts. Second, in a later manuscript, in 1914 (Frege 1969:225–228), he expounded a more sophisticated account of the role of definitions in clarifying concepts which more adequately meets the problem raised by the examples we have just considered. This theory takes a more pragmatic view of definitions by treating them as explications (in the sense of Quine and Carnap), and bypasses the question of the meaning relations between the term in its pre- and postdefinitional uses.

Concerning (3), Frege should, I believe, also have expanded upon his reply to Husserl by pointing out that his analysis of number contains an analysis of the form of number statements, which is independent of his particular definitions. His conclusion that ascriptions of number attribute a property to a concept attains a level of logical sophistication and refinement nowhere touched by Husserl's psychological investigations.

Finally, with respect to (2), we shall later see how Frege complemented his theory of definitions with an account of the introduction of primitive terms. The meanings are to be explained by the use of extrasystemic discussions in which metaphors, models, and even rough formulations of the main points of the theory being introduced are permitted. With this doctrine, Frege makes explicit the method used by scientists to communicate new theories. The doctrine did not receive explicit formulation until the beginning of the twentieth century, when Frege and Hilbert became embroiled in a controversy about the nature of so-called implicit definitions of primitive terms via axioms. There are, nonetheless, foreshadowings of this doctrine in some of Frege's writings that antedate the review of Husserl's book. I have in mind Frege's statements that his notions of *function, concept,* and *object* are absolutely primitive notions which cannot be explained by definitions, so "there is nothing for it but to lead the reader or hearer, by means of hints, to understand the words as is intended" (Frege 1892b:193; see also p. 204).

Frege could have protested—and should have done so—against Husserl's method for dealing with undefinables. For the method of tracing the psychological origins of primitive concepts would be of use only in achieving a mutual understanding of the expressions signifying them if an ideation theory of meaning or of concepts is correct. For even if such an investigation produces rules for the production of certain ideas or mental images, there is no guarantee, unless an ideational theory is correct, that someone who employs these rules will acquire the appropriate concept. Furthermore, the practice that scientists have successfully employed in attaining an adequate grasp of the primitives of their theories in no way resembles the method implicit in Husserl's account. Thus, Frege but not Husserl has a ready (partial) explanation for their success.

Our discussion has shown that there can be a variety of motives for tracing the psychological origins of our mathematical notions—those of Husserl are quite different from Locke's. Still, history has borne out Frege's judgment that such investigations are valueless to mathematics itself. Furthermore, there can be little doubt that his achievements in formal logic turned the foundations of mathematics and logic away from the psychological direction in which they seemed to be headed during the high point of his career. On the other hand, we should not let Husserl's psychologism detract from the deep points concerning the role of definitions which his critique of Frege unearthed. But more on that later.

The Treatment of Logic as the Science of Human Thinking; the Reduction of Truth to Acceptance

In his *Grundgesetze* and in two recently published manuscripts, both entitled "Logik" (one written between 1879 and 1891, the other in 1897) (Frege 1969), Frege mounted his more thorough attacks against the psychologistic approach to logic itself. Besides the antisubjectivistic arguments which have already been reviewed, these works include a critique of the reduction of truth to acceptance as true and the identification of the laws of thought with general regularities describing human thought processes, both of which would convert logic, in his opinion, into branches of psychology and sociology. In this section I shall discuss Frege's attack on these psychologistic themes while maintaining a separation of them from his attacks on subjectivism or ideational theories, and then I shall examine the bearing of this attack upon the work of Christoph Sigwart.

The basic element of Frege's critique is the assertion of the independence of truth from human judgment:

> Being true is different from being taken to be true, whether by one or many or everybody, and in no case is it to be reduced to it. There is no contradiction in something's being true which everybody takes to be false. I understand by "laws of logic" not psychological laws of takings-to-be-true, but laws of truth. If it is true that I am writing this in my chamber on the 13th of July, 1893, while the wind howls out-of-doors, then it remains true even if all men should subsequently take it to be false. If being true is thus independent of being acknowledged by somebody or other, then the laws of truth are not psychological laws: they are boundary stones set in an eternal foundation, which our thought can overflow, but never displace. It is because of this that they have authority for our thought if it would attain that truth. They do not bear the relation to thought that the laws of grammar bear to language; they do not make explicit the nature of our human thinking and change as it changes. [Frege 1893:I, p. XVI]

We need to supply an argument from Frege's distinction between truth and acceptance as true to the conclusion that logic is not concerned with regularities describing what we take to be true. Presumably, the connection is to be found in the fact that logic is concerned with logically valid or truth-preserving inferences. Thus, if truth is independent of acceptance, so must be logical validity; for otherwise a truth-preserving inference form could be counted as invalid merely because we reject the inferences it countenances. The connection I am attributing to Frege between truth and inference is borne out by this passage:

> Logic has only to do with those grounds for a judgment which are truths. Judgments, in which one is conscious of other truths as already justified, are called *inferences*. There are laws concerning this type of justification, and the goal of logic is to set forth these laws of correct inference.
>
> ... It would not be incorrect to say that the logical laws are nothing other than a development of the content of the word "true." [Frege 1969:3]

Frege also argued effectively against psychologism by pointing out that its conception of the laws of logic led to the absurd consequence that these laws would vary as human thought patterns underwent evolutionary changes and, accordingly, should, in their exact formulations, contain explicit references to the human beings for which they are valid:

> In our time, inasmuch as evolutionary theory has marched victoriously through the sciences and the historical approach threatens to infringe its proper boundaries in all matters, one must deal with questions of an astonishing sort. Given that humans like all living beings have evolved and will evolve further, have the laws of their thought always obtained and will they always obtain? Will an inference which is now correct be correct in a thousand years and was it correct a thousand years ago? Obviously here lies a confusion between the laws of actual thought and those of correct inference. ... Laws in the sense of natural laws, psychological, mathematical or logical laws cannot change at all if taken exactly. [Frege 1969:4]

All determinations of the place, the time, and the like, belong to the thought whose truth is in point; its truth itself is independent of place or time. How, then, is the Principle of Identity really to be read? Like this, for instance: "It is impossible for people in the year 1893 to acknowledge an object as being different from itself"? Or like this: "Every object is identical with itself"? The former law concerns human beings and contains a temporal reference; in the latter there is no talk either of human beings or of time. The latter is a law of truth, the former a law of people's taking-to-be-true. . . . These mixings-together of wholly different things are to blame for the frightful unclarity that we encounter among the psychological logicians. [Frege 1893:I, p. XVII]

I am not aware of anyone who held exactly the views of logic and truth which Frege criticized. Benno Erdmann had them foisted on him in the *Grundgesetze,* but I have been unable to subject his works to further examination. Still we can achieve an idea of why one might hold such views by examining Christoph Sigwart's *Logic*. His criterion of truth is acceptance-based, and he considered the investigation of actual thought processes to be part of the task of logic. In addition to this, Frege must have been cognizant of Sigwart's work, since Husserl discusses it in several places in his *Philosophie der Arithmetik.* Unlike Erdmann, Sigwart was an important and influential philosopher during Frege's time.

Sigwart separated logic from psychology—the latter being concerned with "thought as it actually is," the former being concerned "to conduct thought in such a manner that the judgments may be *true*" (Sigwart:I, p. 9). In its striving for truth, logic is normative—the "Ethics of Thought" (Sigwart:I, p. 20). While Frege would applaud all this, in the very same passages Sigwart reduces truth to acceptance:

The judgments may be *true*—that is, necessary and certain— that is, accompanied by a consciousness of their necessity and therefore universally valid. [Sigwart:I, p. 9]

The surrounding passages—which will be quoted below—make it clear that "universally valid" here means accepted by all thinking beings. Sigwart's reduction of truth to universal validity came about in the following way. We cannot compare our thoughts about the external world with it but (if it exists) they must be the same for all knowing subjects; "thus thought which knows the Existent is of necessity a universally valid thought" (Sigwart:I, p. 7). And if there is no external world, "it still remains true that the ideas to which we attribute objectivity are those which we produce with a consciousness of necessity. The fact that we regard anything as existing implies that all other thinking creatures of like nature with ourselves (even when only hypothetically assumed) would also be forced with the same necessity to regard it as existing" (Sigwart:I, p. 8). This approach to truth permits us to remain aloof from the dispute between subjective idealism and realism (Sigwart:I, p. 308). The argument here is, if I understand Sigwart correctly, that whether or not the external world exists, the judgments which we call "true" and which purport to describe it should be in conformity with those of others. We do not have direct access to anything but our own mental contents, but any thoughts that are respondent to something objective would appear to us to be caused by it and independent of our will, that is, psychologically necessary for us and for anyone of a similar mental constitution. Thus, this psychological necessity and universal agreement may function as a criterion of truth.

Sigwart is thus attempting to develop a concept of truth which will be independent of the epistemological and metaphysical realism that underlies Frege's conception of truth. We can see also that this attempt has miscarried; since, on the one hand, it will count a falsehood as true if everyone fervently takes it to be true; and, on the other hand, its reliable applications (if there are any) are restricted to truths that can be known a priori. Indeed, the mere existence of a dispute concerning a proposition would seem to disqualify it from being true. And if some doubt it while others doubt its negation, then neither can be true! In typical Fregean fashion we seem to have thrown Sigwart into a subjectist morass.

Sigwart would respond, I believe, by arguing that while in ordinary life we do encounter a great lack of agreement concerning matters of fact, this would be rectified if we disciplined our thinking by following the precepts of logic. Since much of the discordance we observe is due to variations in linguistic usage, logic must concern itself with linguistic reformation:

[*51*]

We cannot speak of the complete logical certainty of a judgment nor of its invariable validity unless we assume that whenever a judgment seems to be the same as another, because clothed in the same language, it is really the same; that the same statement is made about the same subject. And before we can say that any given judgment is universally valid—*in concreto,* therefore, comprehensible and convincing for all—we must assume that the ideas contained in it are common to all and the same for all. The anarchy of natural thought is completely excluded by the ideal of perfect thought, and a logic which is to contain the rules of perfect thought must begin by determining what are the conditions to be fulfilled by ideas themselves as elements of the judgments.

. . . [these] ideas which enter into the judgment as subject or predicate should be absolutely constant, completely determined, the same for everyone, and denoted by unambiguous terms. [Sigwart:I, p. 243]

To achieve this ideal language, Sigwart claimed, we must analyze our ideas into their ultimate constituents and bring it about that everyone carries out this analysis and the synthesis of new ideas in the same way (Sigwart:I, pp. 257–258). For this we must study our actual thought processes:

This analysis could not be completely carried out unless based upon a perfect knowledge of the laws which govern the formation of ideas, and such a knowledge alone could assure us that the elements were the same for all thinking beings. [Sigwart:I, p. 254]

So once again we find ourselves directed to carry out the sort of psychological investigation that we found in the work of Locke and Husserl. Obviously, Frege's remarks about the irrelevance of this to logic and the foundations of mathematics also apply here.

Looking back over the discussion by Sigwart, we see that he was attempting to lay down foundations for logic which are independent of metaphysical assumptions about the external world. Yet his epistemological premise—that we have access only to our own ideas and thoughts—almost necessitated his psychologistic conclusions. And that in turn opened him to Frege's critique.

Psychologism has proved to be an unfruitful approach to logic. Frege's formal logic and his semantics have done much toward demonstrating this—probably far more so than the criticisms we have reviewed. These criticisms were quite successful in exposing unclarities in the particular works toward which they were directed, and they raised problems with which any antirealist philosophy of logic must grapple. Yet I hesitate to attribute a definitive refutation of psychologism to Frege because much of his attack ultimately came down to laying out very clearly and persuasively his opposing view.

Formalism: Thomae, Curry, and Hilbert

While Frege's philosophical colleagues tended to look inward to the human mind as the source of mathematical existence and knowledge, several of his fellow mathematicians sought this source instead in the mathematical symbolism or formalism itself. Although their attempts were poorly executed and, as a result, thoroughly ridiculed and devastated by Frege, their general line of thought, which has become known as *formalism,* has been a fruitful approach to the foundations of mathematics. For this reason, I shall also examine the formalistic views of Curry and the later Hilbert. Neither was discussed directly by Frege, but they can be seen as natural responses to his critiques of his colleagues' theories.

I find it useful to distinguish at least three varieties of formalism: (1) *game formalism,* which takes mathematics to be a meaningless, chesslike game in which the symbolism functions as the "board and pieces"; (2) *theory formalism,* which treats mathematics as the theory of formal systems; (3) *finitism,* which views *part* of mathematics as a meaningful theory of certain symbolic objects and the remainder as an instrumentalistic extension of the former. Game formalism and theory formalism trace to Frege's colleagues, Heine and Thomae, although Frege himself presented the first clear formulations of these views. Finitism originated with Hilbert. In a later chapter I shall discuss a close relative of formalism, *deductivism,* which emerged in the course of the controversy between Frege and Hilbert concerning the foundations of geometry.

[54]

Game Formalism

Formalism has a respectable mathematical history dating back at least to the introduction of the imaginary numbers, by Bombelli in the sixteenth century, to furnish numerical roots for previously unsolvable equations such as "$x^2 = -1$." Because Bombelli believed merely that he was introducing symbols, treating them as if they designated numbers and subjecting them to algebraic operations without actually producing the numbers themselves, he reflected his many reservations about this feat by calling his numbers *imaginary numbers*. For many years, mathematicians operated with this number system and studied its algebraic properties while retaining their qualms concerning its reality. Only after the geometric representation of the imaginary and complex numbers offered by Gauss in the early nineteenth century became well known did mathematicians begin to concede their existence. Yet this meant that at that time both analysis and the theory of complex numbers were based upon geometric foundations. During the last half of the nineteenth century, Weierstrass, Dedekind, and Cantor set about to remedy the situation by arithmetizing analysis and the theory of complex number. To do this, they introduced infinite processes (i.e., limits, infinite sequences, and infinite series), through which may be defined derivatives and the irrational, transcendental, and complex numbers. However, Kronecker, a mathematician of great influence during his day, objected vehemently to infinite processes as "metaphysical" and argued that only the natural numbers and entities finitely constructible from them should be admitted in mathematics. Heine and Thomae were among those contemporaries of Kronecker and Frege who reacted to Kronecker's criticism in the spirit of an unabashed Bombelli. I will quote Heine:

> Suppose that I am not satisfied to have nothing but positive rational numbers. I do not answer the question "What is a number?" by defining number conceptually, say by introducing irrationals as limits, whose *existence* is presupposed. I define from the standpoint of the pure formalist and *call certain tangible signs numbers*. Thus the existence of these numbers is not in question. [Quoted by Frege 1903a:sec. 87]

[55]

According to Heine, then, we have no more reason to doubt the existence of a number—once a symbol for it is given—than we have to doubt the existence of the symbol, since the symbol and the number are one and the same. Yet one cannot call numbers symbols and leave it at that. As Frege was quick to point out, written symbols have properties that we don't ordinarily ascribe to numbers. For example, written symbols have physical and chemical properties. Numbers do not. On the other hand, numbers have properties which—at least at first sight—cannot be ascribed sensibly to symbols. We don't talk of a certain symbol on the blackboard as being greater than another symbol on the blackboard, unless by that we mean that it is in fact physically larger than that symbol. Yet we can write down a small token of "3" and write a large token of "2" and still assert that 3 is greater than 2 (Frege 1885:97-98).

Another contemporary of Frege's, Thomae, answered this objection by avoiding Heine's explicit fusion of number and numeral. According to Thomae, although arithmetic deals with nothing but numerical symbols, it is not a theory of these symbols; rather it is a game played with them.

> The formal conception of numbers accepts more modest limitations than does the logical conception. It does not ask what numbers are and what they do, but rather what is demanded of them in arithmetic. For the formalist arithmetic is a game with signs which are called empty. That means that they have no other content (in the calculating game) than that they are assigned by their behavior with respect to certain rules of combination (rules of the game). The chess player makes similar use of his pieces; he assigns them certain properties determining their behavior in the game and the pieces are only external signs of this behavior. To be sure, there is an important difference between arithmetic and chess. The rules of chess are arbitrary, the system of rules for arithmetic is such that by simple axioms the numbers can be referred to perceptual manifolds and can thus make important contributions to our knowledge of nature. [Quoted by Frege 1903a:sec. 88]

[56]

Thus, the physical and chemical properties of written numerical signs are no more relevant to the study of arithmetic or the playing of the arithmetic game than the corresponding properties of chessmen are to the playing of chess. (Of course, some physical properties are relevant. Chessmen cannot be made out of whipped cream, and numerical signs cannot be written in invisible ink. But we tend to feel that these are simply practical considerations and not of theoretical importance.)

It must be admitted that Thomae's view has a lot of initial plausibility, for a great deal of practical mathematics, at least, consists of symbol manipulation without particular emphasis upon the meaning of the symbols. We do learn the multiplication table by rote and often simply manipulate symbols as we make our calculations. Although there is much more to mathematics than calculations, all of mathematics does involve symbols. Thus, it seems that one might make a case that all mathematics is nothing but symbol manipulation according to preset rules. The questions are whether Thomae did in fact make out such a case; and if it can be made, does that give us an adequate philosophy of mathematics? Even Frege, it will turn out, granted that one can view mathematics as mere symbol manipulation. But he was convinced that this does not give a true account of the nature of mathematics (Frege 1903a:sec. 90).

The early formalists did not even succeed in showing that mathematics could indeed be cast as a game of symbols. It was Frege who pointed out the inadequacies in their view and indicated how those inadequacies could be corrected. Thomae wrote, for example: "The system of signs of this computing game is constructed in the familiar way from the signs 0, 1, 2, 3, 4, 5, 6, 7, 8, 9" (Thomae:434). Frege observed that Thomae used many more signs than he listed, and gave no indication of how these additional signs are composed from the 10 signs given. In other words, Thomae's formation rules are completely unspecified, and the only reference he makes to them is to our understanding them in "the familiar way" (Frege 1908). It is legitimate for a mathematical logician writing today to say that the formation rules for a system are to be understood tacitly in "the familiar way" while omitting them in a particular paper, for every competent mathematical logician knows how to supply them. But Thomae was proposing to set up arithmetic as a

[57]

symbolic game for the first time; therefore, there was no "familiar way" of setting it up.

Frege also accused Thomae of presupposing meaningful arithmetic in his formal arithmetic. In addition to formation rules, telling the players which symbols are pieces in the game, Thomae needed transformation rules that state how to manipulate these symbols. Here, again, he relied upon meaningful arithmetic in stating them. He wrote, for example, that one of the rules of his system is $a + a' = a' + a$ (i.e., the usual commutative law for addition). But then Frege had only to ask: What is he supposed to mean here? Certainly, he cannot be asserting that the *symbol* "3 + 1" is identical with the *symbol* "1 + 3." By attributing meaning to these symbols we can make sense of this equality, because we can interpret it to mean that the number obtained by adding 3 and 1 is the same as the number obtained by adding 1 and 3. Yet this interpretation is not open to Thomae. So what can it mean? After probing along these lines, Frege determined that it must mean that the two symbols are interchangeable in any other symbolic context in which they occur. Thomae's basic equations are actually rewrite rules (Frege 1903a:secs. 106–109; 1906b).

Following this line, we can see how game formalism could go. To set up the formation rules, let us use the unary number system, in which symbol "1" is the only initial symbol and the operation of the juxtaposition is used to form additional numerals. Then we introduce decimal notations as abbreviations for these symbols, so that "11" is abbreviated as "2," "111" is abbreviated as "3," and so on. Next, we introduce recursive functions on the natural numbers by viewing the recursive equations as rewrite rules. Having gotten this far, we may take one step further and introduce ratios as composite signs consisting of a decimal numeral written over another one, and introduce an equivalence relation between ratios by taking the ratio n/m to be equivalent to the ratio p/q if and only if mp rewrites as nq. We then may use this equivalence relation to define a rewrite (i.e., an "identity" relation) between equivalent ratios so that any two equivalent ratios may be replaced by each other. Of course, we must not allow division by 0, but this can be avoided by excluding n/p as malformed when p can be written as "0."

Notice that we have introduced numerical symbols of various kinds but have not yet introduced any numerical statements. Thus, we have reflected only the use of numbers in calculation, not their use in theoretical mathematics. Statements of theoretical mathematics will have to be viewed as statements about the symbolism we have introduced so far. Identities can be thought of as stating that two mathematical symbols can be rewritten in terms of each other. A statement of the form $a > b$ would tell us that when a and b are converted to strings of 1's, the string obtained from a is shorter than the one obtained from b. Moreover, we could introduce variables ranging over mathematical symbols and proceed accordingly to form general laws about the mathematical symbolism.

It should also be noted that, strictly speaking, the laws of symbolism are not laws in a metatheory, because there is no object *theory* for the metatheory to treat. The so-called object language contains only names or numerical symbols. It is only when we start talking about symbolism that we start to formulate any theory at all. It is true, of course, that the truths of the so-called metatheory are grounded in the rules of the symbol manipulation that the metatheory is about. But one who operated only at the level of object symbolism would never express such truths or even need to know them. Similarly, children may calculate prodigiously without being aware of the associative and commutative laws of addition and multiplication. Thomae's game formalism suffered because of his failure to recognize this distinction between the game of mathematics and its theory. (Cf. Frege 1903a:secs. 107–109.)

The rewrite-rule approach to arithmetic works for the rational numbers but breaks down when we try to extend it to construct the real numbers in terms of Dedekind cuts or infinite sequences or series. Thomae and Heine introduced infinite sequences of symbols without further ado, so it was easy for Frege to make hash of their attempts. His first point was that there are no infinite sequences of written symbols because no one ever has or ever will inscribe one (Frege 1903a:sec. 124). Thomae used a more sophisticated approach, however, by calling a sequence infinite if it is defined by a rule according to which it *can* be continued indefinitely. So Frege attacked his use of possibility in this definition by arguing that either no sequence is infinite on this definition because it is not possible for humans to inscribe symbols indefinitely or, if a broader notion of possibility is invoked, then all sequences can be prolonged indefinitely. Thus every sequence is finite or every sequence is infinite (Frege 1903a:sec. 125). The rules Thomae actually used were criticized by Frege as being inadequate to the desired prolongations and as tacitly presupposing meaningful arithmetic (Frege 1903a:sec. 129–131).

Frege's harsh criticisms were warranted by the sloppiness of Thomae's exposition, but more can be said for the basic approach. To resuscitate it, let us first abandon the use of written inscriptions in favor of abstract symbols (types, not tokens). Next, let us identify sequences with their rules of generation. Then it is no longer the case that an initial segment of a sequence is indistinguishable from the finite sequence that terminates with the last member of the segment and from the infinitely many infinite sequences that begin with the same segment. We have also parried Frege's objection that either all sequences are infinite or all sequences are finite, since a finite sequence will be one whose rule entails that the generation of the sequence will terminate. Furthermore, we can circumvent Frege's charge of presupposing arithmetic if we formulate our rules carefully. For example, the rule for constructing the sequence "0," "0," "0," "0,"... could be: inscribe a "0" and write a "0" to the right of every "0" that has been inscribed.

Yet, even this procedure fails to produce a classical continuum of real numbers unless we are prepared to admit arbitrary rules for sequence generation. For the classical continuum of real numbers is nondenumerably infinite. Thus, nondenumerably many rules will be required for its construction, and this in turn necessitates that no restriction be placed upon the notation in which the rules are formulated. Yet now we will be committed to the existence of arbitrary rules to obtain surrogates for arbitrary infinite sequences or sets. Surely this is just as "metaphysical" a commitment as that which Thomae and Heine's formalism sought to avoid! Similar difficulties arise if, instead of abstract symbol types, we use possible symbol inscriptions and modal logic; but then the real problem would be that of obtaining a rich enough set of sequences. The "metaphysical" problems of arbitrary rules or possible worlds can be passed over by adopting the standpoint of intuitionism and using free-choice sequences. But, of course, this does not meet the formalist's goal of founding classical mathematics (Heyting).

In sum, Frege's criticisms of Heine and Thomae show in great detail the defects on the execution of their programs. Not only do they repeatedly confuse signs with what they signify, but they also conflate mathematics qua game and mathematics qua theory of a game. By presupposing meaningful mathematics, they fail to present the formation and transformation rules which are crucial to their approach and fail to see the failure of their method to extend to the construction of infinite sequences and sets. To Frege we owe the first explicit formulations of the use-mention and object-metalanguage distinctions and the first clear indications of how a rigorous formalistic philosophy should be presented. Yet even when Thomae's formalism is properly formulated, it founders on the problem of the infinite.

There is another type of game formalism which overcomes this last objection. I will call this *derivation game formalism* to contrast it with the *computation game formalism* of Thomae. According to this type of formalism, mathematics is a game or set of games whose pieces are well-formed formulas of formal systems and whose moves are the transformation rules of these systems. The object of the game or games is to construct derivations in formal systems.

In setting up a formal system, we do need potential infinities of symbols, but we do not require arbitrary sequences. Thus, the problems with the real numbers which plagued computational game formalism do not touch derivation game formalism. The theories of real numbers and infinite sequences will be certain formal systems, and they are no more problematic than is the theory of natural numbers. Given the formalization of current mathematical practice, no technical difficulties seem to impede this brand of formalism.

Frege was well aware that his own system could be treated as a derivation game. He even emphasized that a completely rigorous derivation of mathematics from logic requires that the formal system have this property. Otherwise, one cannot mechanically check proofs to see whether they are correct (Frege 1884:sec. 90–91; 1903a:sec. 90). But just because one can treat mathematics in this way, it does not follow that derivation game formalism is a philosophically satisfactory account of mathematics. This point should be underscored, because philosophers of mathematics often have taken the success of a certain technical program as validating their philosophical claims about the nature of mathematics. It is tempting to take the success that mathematical logicians have had in formalizing extant branches of mathematics as showing that there really is nothing to mathematics except the manipulation of formal systems. But this is at most a necessary condition for the formalistic thesis, which itself requires further arguments.

In addition to technical objections to formalism, Frege had this general philosophical objection:

> An arithmetic with no thought as its content will also be without possibility of application. Why can no application be made of a configuration of chess pieces? Obviously, because it expresses no thought. If it did so and every chess move conforming to the rules corresponded to a transition from one thought to another, applications of chess would be conceivable. Why can arithmetical equations be applied? Only because they express thoughts. How could we possibly apply an equation which expressed nothing and was nothing more than a group of figures, to be transformed into another group of figures in accordance with certain rules? Now it is applicability alone which elevates arithmetic from a game to the rank of a science. So applicability necessarily belongs to it. Is it good, then, to exclude from arithmetic what it needs in order to be a science? [Frege 1903a:sec. 91]

In brief, arithmetic can be applied, whereas meaningless games cannot; so arithmetic is not a meaningless game.

Taken literally, this objection is wrong. Games can be used as training devices or models of more complex real-life phenomena, for example. Frege had inferential applications in mind, however, as his words "a transition from one thought to another" indicate. Suppose, for example, that to cover a floor 10 feet by 6 feet with tiles that measure 1 foot square we need to determine the area of the floor in square feet. Then we might reason as follows: (1) area in square feet = length in feet times width in feet; (2) length in feet = 10; (3) width in feet = 6; (4) 10 × 6 = 60; thus (5) area in square feet = 60. On Frege's view, only true thoughts—not the sentences expressing them—can be premises of inferences. (See Chapter Five for Frege's views on inferences and their premises.) Thus, premise (4) cannot be an empty formula but must express a true thought; and to express a *truth* its singular terms must, on Frege's account, refer to objects. In short, the application of arithmetic in an inference necessitates that it be both meaningful and referential.

This objection cuts deep. For it obviously applies not only to mathematics but to other sciences as well, thereby casting doubt upon instrumentalism and fictionalism in physics as well. Furthermore, since a theory of meaning that treats all scientific discourse uniformly is so much simpler than one which draws distinctions between the sciences or within the sciences, the objection places the burden upon the instrumentalist. How, for example, is mixed discourse such as our argument, or mixed sentences such as (2), (3), and (5), to be handled?

I shall not consider the details of the many ways of responding to this objection, which are in the literature. Most are not relevant to the Frege-Thomae discussion. For example, the response that the inference in question might be eliminated in favor of one that does not refer to numbers via, for example, Craig's Theorem, would seem to be beside the point, since both Frege and his opponents wished to retain arithmetic and its applications in their usual form. Two responses seem appropriate to the present discussion: (1) arithmetic is a calculating device and need be no more meaningful than an adding machine; and (2) arithmetic, although meaningless in itself, receives meaning via its applications.

Frege did not take up the first claim, perhaps with justification. There are serious dis-analogies between our use of arithmetic and calculators. Calculators assist us in finding true arithmetical equations. The truth of arithmetic is presupposed in the construction of equations, the explanation of their behavior, and the justification of their reliability. When a calculator is used in drawing conclusions, neither it nor its output figures as premises in inferences. Instead, statements describing the inputs and outputs of the calculators are used. For example, if we employed a calculator in our tile example, we would infer that 10 × 6 = 60 from the fact that it gave the output "60" for the inputs "10," "6," and "×."

Thus, if the equation "6 × 10 = 60" is to be regarded as an outward sign of a move in the "calculating game," our inference, strictly speaking, should have a description of our calculation in place of (4). Then we would need to know the relevance of this to the conclusion that 60 tiles are required. Perhaps we could view the calculations as furnishing us with information about the results we would obtain from alternative methods of counting the tiles (e.g., counting them in six rows of 10 each versus counting the entire area at once). But we would still require an explanation of why the game works, and here meaningful arithmetic seems unavoidable.

Frege did consider the second response but rejected it on the grounds that arithmetic is too general a science to be founded upon its applications. If it were based upon a single application—say through a geometric definition of number—then its other applications might be precluded (Frege 1884:sec. 19). On the other hand, arithmetic certainly should not be ambiguous—the symbol "2," for instance, denoting a length in one application and a weight in another. Thus, arithmetic must deal with what is common to all its applications and, to be a science, must have its own subject matter (Frege 1903a:secs. 73, 92).

Another way out of the dilemma Frege poses, which traces to Hilbert, will be examined in a later chapter. This approach treats mathematics as an empty theory whose terms are schematic letters and whose sentences are quantificational schemata. Mathematics, on this view, reduces to pure logic and is concerned solely with deducing logical consequences from arbitrarily selected schemata. In applied mathematics the schematic letters are interpreted via empirical predicates. Mathematics no longer can be interpreted as a game, even if its formulas are not meaningful, because logic rather than arbitrary stipulation becomes the controlling factor. Frege also rejected this approach, by reaffirming the scientific nature of mathematics, with its assertion of the truth of its axioms as well as their logical consequences.

To me, a most telling objection to game formalism is to be found in Frege's protestation that arithmetic is a science, not a game; and that arithmetical formulas can be used to make assertions, raise questions, express hypotheses, and so on, which no uninterpreted configurations of game pieces can (Frege 1903a:sec. 91; 1906a:396). It may be granted

that assertions in mathematics can be construed, incorrectly in my opinion, as assertions that certain formulas are provable, and these in turn as
claims that certain moves in a game of symbol manipulations are possible. But this shows at most that mathematics is a theory of a game—one
that even mathematicians themselves hardly play. Construing mathematics itself as a game only gives rise to the most tortuous explanations
of why it is taken so seriously, why mathematicians appear to make
claims through it, and why it appears to strive for truth. In this respect
the Fregean view of mathematics as a science surely has the advantage.

Curry's Formalism

Haskell Curry's thesis that *mathematics is the science of formal systems* is a contemporary descendant of nineteenth-century formalism
which escapes Frege's objections. Furthermore, by dealing with objects
such as formal systems that can be concretely represented (at least in
part), it appears to offer more promise of leading to an adequate epistemology for mathematics than a Platonist theory such as Frege's.

After some analysis we shall interpret Curry's view as taking the
basic assertions of mathematics to have the form "such and such is a
theorem of such and such a formal system." Although this permits
mathematics to be a meaningful science with genuine truths, it has the
disadvantage of interpreting all of mathematics as what we would count
ordinarily as metamathematics. Since Curry has a somewhat esoteric
conception of formal systems, it will be necessary to consider the details
of his formulation.

According to Curry, a formal system is completely identified by
means of its primitive frame, which consists of three elements: (1) the
terms of the formal system, (2) the elementary propositions of the
formal system, and (3) the elementary theorems of the formal system.
Terms are specified by first presenting a list of primitive terms or, as
Curry calls them, *tokens*. (This is somewhat of a misnomer, since the
terms of a formal system need not be written expressions. In fact, Curry
declines to say what the terms of a formal system are.) Next, one
presents a list of *operations* which combine terms already formed into
new terms. Each operation is specified with the number and kinds of

argument that it takes. Finally, rules of formation are given which specify how terms may be built out of primitive terms using the operations. As an example, suppose that we are given 0 as a primitive term, the prime as an operation of one argument (taking any term as an argument), and this formation rule: If a is a term, then a' is a term. This will then generate the following class of terms: 0, $0'$, $0''$, $0'''$, and so on.

Elementary propositions are specified by first laying down a list of *elementary predicates* with the number and types of argument that each predicate takes. An elementary proposition is then formed by applying such predicates to sequences of terms of the proper number and kind. Thus, if we introduce identity as a primitive two-place predicate, the elementary propositions obtained in our example will be identities between terms (e.g., $0 = 0$, $0 = 0'$, $0' = 0''$).

Elementary theorems are determined by first presenting a list of elementary propositions under the label *axioms*. This list may be finite or infinite, but if it is infinite it must be given effectively. No special attribute is required for an elementary proposition to be an axiom; for the axioms are the (arbitrary) choice of the person designing the formal system. Next, a list of *rules of procedure* is given. The rules must be effective in the sense that one can determine effectively whether a given elementary proposition follows from other elementary propositions according to a given rule in the list. To continue our example, we list as an axiom "$0 = 0$," and then lay down as a rule of procedure the stipulation that if $a = b$ is an elementary theorem, so is $a' = b'$. The *elementary theorems* of a formal system are simply the axioms and the elementary propositions that can be generated from them according to the rules of procedure. Thus, the elementary theorems of the formal system that we have given as an example will be all the elementary propositions of the form $a = a$.

Although Curry uses the words "term," "token," "predicate," "proposition," he does not identify a formal system with a system of linguistic entities. When a formal system is introduced, a symbolism will be required to present it. Symbols will be needed to list the tokens and to present the operations and predicates, and symbolic constructions will be required to symbolize the construction of terms. Similarly, sentences will be needed to symbolize elementary propositions. However, the formal system is not to be identified with the symbol system used to present it. Of the many ways to represent a formal system, one is the symbolism used to present it. Here we think of its terms as singular terms and its elementary propositions as sentences. But, according to Curry, we could just as well think of terms as buttons and of operations as ways of stringing buttons together. Or we could take terms even as some kind of abstract or mental entities. All of this is irrelevant, he believes, to the study of formal systems and mathematics. Curry concedes that one is very unlikely to concoct a formal system without having in mind some representation for it, but he does not consider this to be important (Curry:8–27).

A major reason Curry gives for his claim that the theory of formal systems requires no ontology is to be found in his characterization of truth for elementary propositions. He asserts that the truths of a formal system are its elementary theorems, yet he denies that he is using a nonstandard definition of truth or identifying it with theoremhood. Although I find his account obscure at this point, I think his reasoning is as follows. The extension of a predicate P of a formal system S is defined by the condition that the ordered n-tuple of terms $\langle t_1, \ldots, t_n \rangle$ belongs to the extension of P just in case the elementary proposition $Pt_1, \ldots, t_n$ is a theorem of S. The elementary proposition $Pt_1 \ldots t_n$ is true, however, just in case $\langle t_1 \ldots t_n \rangle$ belongs to the extension of P. (This is the usual Tarskian notion of truth.) Thus, although theoremhood is both necessary and sufficient for truth in the case of an *elementary proposition,* it is not identical with it, since truth has a broader extension in other cases. Of course, Curry succeeds in retaining the usual characterization of truth, while cutting it loose from reference and ontology, only by using an unorthodox approach to extensions.

This trick can be easily extended to formalizations of advanced mathematical theories by using elementary formal systems whose only *predicate* is the assertion sign. Suppose that A is a well-formed formula or sentence of an ordinary formal system. Then in the Curry correlate of this system the formula A will be matched by a complex *term* whose symbolic representation is also A. The elementary propositions of the Curry correlate will have the form $\vdash A$ and will be true just in case A is a theorem of the original system. For example, in the ordinary formal

[67]

system ZF the following *sentence* is a theorem $(x)\ (\exists y)\ (x \in y)$. In the Curry correlate, ZF_c, the expression "$(x)\ (\exists y)\ (x \in y)$" symbolizes a *term* and the expression "$\vdash (x)\ (\exists y)\ (x \in y)$" symbolizes a true elementary proposition. In this way every elementary proposition can be viewed as a statement to the effect that a given formula is a theorem of some (non-Curry) formal system (Curry:28–30, 37–38).

This method of interpreting Curry's formal systems shows how the science of formal systems can be related to mathematics as we usually think of it. For suppose that every branch of (ordinary) mathematics is formalized through a formal system, not in Curry's sense but in the usual sense of mathematical logic. Curry can then be interpreted as claiming that the elementary propositions of (his) mathematics state that certain sentences of these formal systems are theorems. Hence, in effect, Curry has made mathematics into metamathematics.

Thus far the metamathematical propositions that Curry recognizes are rather restricted. But Curry allows us to enrich metamathematics by forming *metapropositions* by compounding and quantifying elementary propositions. Then we can recognize, in addition to elementary propositions, metapropositions such as the following: If the axiom of choice is added as an axiom to the ordinary formal system for Zemelo-Frankel set theory, then the well-ordering theorem becomes a theorem of that formal system. If at least one sentence is not a theorem of this formal system, and if we add the axiom of choice as a new axiom, at least one sentence will not be a theorem of the augmented system. Curry claims that this will allow us to formulate the usual results of mathematical logic (Curry: 50–53). It is not clear to me, however, that we can achieve formulations of model theoretic results, since Curry has no concept of interpretation or satisfaction that can be expressed via elementary propositions.

Curry distinguishes between propositions about elementary formal systems which can be proved constructively and those which involve "extraneous considerations or infinitistic assumptions such as the semantic investigations of Tarski and Gödel's proof of the completeness of the elementary calculus of predicates" (Curry:53). He admits that the distinction between constructive and nonconstructive metapropositions is rather vague. Nonetheless, it is a critical distinction for him because he claims that a metaproposition whose proof is nonconstructive must not be considered as a purely mathematical proposition.

The last consideration would exclude all nonconstructive propositions from the domain of mathematics proper. For the truth of these propositions depends on idealistic assumptions of one form or another which do not arise in the case of the constructive propositions. The former are therefore mixtures of mathematics and something else. This does not mean that they do not have mathematical interest. Neither mathematics nor any other science can develop in complete isolation. As a matter of fact the mathematics which is the usual concern of working mathematicians is rarely pure. . . . Thus the propositions which arise from a nonconstructive argument are not to be dismissed as per se meaningless. They have at least a heuristic value; they show what may result if something more is assumed; and in many cases they may be formalized and so give rise to purely mathematical propositions at a higher level.

In a footnote to this passage, Curry adds the following remark:

This is the case with ordinary mathematical analysis. If one thinks of the theorems of analysis as metatheorems—which is what the ordinary genetic treatment amounts to—then these theorems are non-constructive; but we can formalize them, i.e. we can find formal systems which include (in a certain sense) all of mathematical analysis. Thus the above definition does not exclude anything which is mathematics as usually understood. [Curry:56–57]

This is enough of Curry's view. Let us now consider the advantages of his view over those of the early formalists. The most notable advance is the precision with which he formulates his theory. He no longer relies upon meaningful mathematics, as Frege would call it, to set up his formal systems. Moreover, his formal systems are not games; they are systems of propositions that are either true or false; thus, Curry avoids the charge of depriving mathematics of meaning. Curry also effects some ontological economy by recasting mathematics in terms of formal systems. We no longer have to recognize arbitrary sets, sequences, functions, or real numbers as the elements of mathematics. Curry neatly accounts for the objectivity of mathematics by saying that although individual mathematicians introduce formal systems by defining their primitive frames, once a formal system has been introduced its properties are completely determined independently of further human activity. A further advantage is that at least for a large class of mathematical propositions, Curry has a completely clear and definite criterion of truth—provability according to certain specified rules from certain specified axioms.

Curry's view is further buttressed by the existence of alternative geometries and set theories. Instead of wondering whether one of these geometries or set theories is true and the others false, Curry can simply say that we have here different formal systems, each of which can be studied in its own right. Indeed, one can even study inconsistent formal systems if it fulfills one's purposes. In this connection, Curry distinguishes mathematical truth from the acceptability of a given formal system for a given purpose. He would not assert, for example, that Zermelo-Frankel set theory is mathematically true, since only elementary and metapropositions are mathematically true. On the other hand, he could decide that Zermelo-Frankel set theory qua formal system is more suitable for formalizing mathematics than, for example, the simple theory of types qua formal system.

Curry also can handle Gödel's Theorem, which states (roughly) that any consistent formal system adequate for elementary number theorem contains sentences that are neither provable nor refutable through proofs of the formal system in question. Some philosophers have argued that this theorem shows that mathematical truth outruns proof relative to a given formal system, and they have invoked it against formalism. Curry can argue that Gödel's result simply shows that the extension of the theoremhood predicate does not include certain terms, and that this fact is just another truth about the formal system in question. Since he does not believe in the informal concept of mathematical truth, he can interpret the other limitative theorems of formalization similarly, as not affecting his position.

Turning now to the criticism of Curry's theory, it is most striking how it excludes almost all of mathematics, except for metamathematical investigations. The results of ordinary mathematics can be accommodated within Curry's formal systems, to be sure, but ordinary mathematical *practice* is degraded to the status of a "heuristic." Yet most mathematicians, even those who are formalists, do not reason *in mathematics* along Curry's lines by referring directly to the rules of a formal system. Instead, they make use of the intuitive properties of the logical statement connectives and quantifiers. Not only does Curry's account refrain from explaining mathematical practice; it also fails to answer questions about formal systems themselves which seem intimately connected with mathematical practice. Why are we interested in some formal systems and not in others? Why must inconsistent systems be modified or repaired? What difference does it make if we use only first-order logic or higher-order logics as well? Set theory or the theory of types? Although these are not easy questions to answer, they do appear to be sensible and worthwhile. But without some intuitive conception of mathematics, it would seem impossible to even begin to answer them. This deficiency in Curry's approach makes the obscurities in his ontology and epistemology all the more exasperating.

Let us make up ontological questions first. Curry thinks these are entirely irrelevant to the study of formal systems and thus to mathematics.

> Although a formal system may be represented in various ways, the theorems derived according to the specifications of the primitive frame remain true without regard to changes in representation. There is, therefore, a sense in which the pimitive frame defines a formal system as a unique object of thought. This does not mean that there is a hypostatized entity called a formal system which exists independently of any representation. . . .
>
> It is unnecessary to inquire further into the meaning of a formal system. It is characteristic of mathematics that it considers only certain essential properties of its objects, regarding others as irrelevant. One of these irrelevant questions is that of the ontology of a formal system. The question of which representation is the real or essential one is a metaphysical matter with which mathematics has no concern. [Curry:30–31]

This passage contains two errors. First, it confuses the ontological commitments of a representation of a formal system, or of an interpretation of a formal system, with the ontological commitment engendered by the theory of formal systems. According to Curry, formal systems are the objects of mathematical study, so that mathematicians do commit themselves at least to the existence of formal systems.

We must agree with Curry, however, that a mathematician can proceed to do mathematics without inquiring into the nature of its ontology. There have been times, of course, when the investigations of mathematicians have taken on an ontological air and, indeed, have prompted ontological doctrines from philosophers. When a mathematician presents a reduction of all of mathematics to a single framework, such as to set theory or category theory, he can be taken as proposing a new ontology for mathematics. But it is not clear that he *must* be so taken. For we can conceive of a mathematician arguing as follows: "It is not at all clear to me what sets are. They are mathematical entities, and they have the same status as any other kind of mathematical entity. I do know the axioms for set theory, and therefore I know the methods that are acceptable for reasoning about sets; and that is all that is necessary for me to do mathematics. I can show you that I can construct numbers from sets, although I need not settle the question as to whether this shows that numbers really are sets. That is not relevant to my mathematical purposes, because in some mathematical context I can just take numbers as my starting points and operate with them. In other mathematical contexts it is more economical or mathematically elegant to start with sets and build up numbers out of them." Thus, it is quite plausible that mathematicians can do mathematics without raising questions concerning the nature of its ontology. However—and this is where Curry makes his second mistake—this does not imply that mathematics or the theory of formal systems has no ontology.

Since Curry has failed to raise the question of the ontology of the theory of formal systems, let us do it for him. Had he confined himself to elementary propositions, it would have been possible to interpret the theory as being ontology-free. But Curry does admit bound variables ranging over formal systems and their elements. Indeed, he even commits himself to infinitely many formal objects. This precludes the identification of formal systems with physical objects or concrete inscriptions—an identification never proposed by Curry in any case. Unfortunately, Curry also closed the other ontological avenues. His criticisms of intuitionism and platonism indicate further that he would accept neither a mental nor an abstract ontology for his theory.

The situation is further clouded by Curry's ambivalence concerning constructivity. For until we know whether his quantifiers are to be read constructively or nonconstructively, we will be barred from determining whether formal systems are potentially or actually infinite objects.

Thus, Curry should retreat to the position wherein the only propositions of mathematics are the elementary propositions of formal systems. The rest of *metamathematics* could be formalized and itself treated as a formal system. This would place the general theory of formal systems in the same position as that proposed by Curry for ordinary mathematics. Then the ontological talk of ordinary mathematicians and metamathematicians would be viewed as a purely heuristic device that allows one to work with formal systems and derive elementary propositions in them. This is a high price to pay for purging mathematics of its ontology.

Curry's epistemology is similarly lacking. He makes two statements concerning the epistemology of a theory of formal systems. The first is that in verifying that a given sequence of elementary propositions is a proof, we need only inspect the sequence and check it for conformity to the rules of procedure. This, he says, is a demonstration *ad oculos* (Curry:31–32). In other words, to check that a given sequence is a proof, we merely look and see. Since it is not necessary to use inference, determining whether something is a proof is completely independent of logic. Curry's second statement is that the essence of mathematics is that it uses recursive definitions and mathematical induction:

> The essence of mathematics is that we make definitions by recursion, and then draw up particular consequences by applying the definition and general consequences by mathematical induction. There is, therefore, a certain amount of justice in the view that mathematical propositions are consequences of definition; but, since the definitions are recursive, mathematics does

[*73*]

not have the trivial character which that seems to imply.... Of
course, there is intuition involved in all this—if one defines
intuition properly the statement is tautological. But the question
of the metaphysical nature of this intuition is irrelevant. To
many of us it seems that the intuition is an empirical, linguistic
phenomenon; but if one wishes to associate with it an a priori
notion of pure time or a mystic contemplation of the absolute, or
other such ideas, the content of mathematics will not be af-
fected. [Curry:57–58]

Although Curry does not say anything more about the epistemology of
mathematics and expressly denies its relevancy, it would seem that
something of a Kantian theory is implicit in his view. For when we
verify that a given elementary proposition is true, we do so by inspect-
ing a proof. Yet the proof is a written inscription. We know that inscriptions are not essential to a formal system, although they are necessary
for human communication. So we could think of the formal system and
its elements as a Kantian concept, and the symbolism and written proofs
could be thought of as exhibitions of these concepts in intuition. Although this might explain Curry's talk about intuition, it would leave
Curry with some very serious epistemological problems and could
hardly be said to have freed mathematics of metaphysical presuppositions.

   Another possibility is that the inspection of proofs is an empirical
process. This seems rather unlikely because then proofs would have to
be written inscriptions, and Curry has no account of how formal systems
are related epistemologically to written inscriptions. In any case, it is
difficult to see how he can claim that the process of verifying a given
inscription of a proof is entirely free of logical reasoning. After all, one
has to check that the proof does conform to the patterns laid down in the
rules setting up the formal system. In the case of short proofs, one might
claim that this is self-evident. Otherwise, one must reason somewhat as
follows: Any inscription of the following form is an axiom. This inscription is of that form; therefore this inscription is an axiom. Any two
inscriptions that are related in the following way are such that one of
them is a direct consequence of the other. These two inscriptions are so

related; therefore, these inscriptions are such that one is a direct consequence of the other. Finally, one would have to reason that inscriptions that are put together in certain ways form a proof, and then conclude that this inscription forms a proof.

Remaining is that nagging question of the distinction between constructive and nonconstructive reasoning, about which Curry has nothing to say. But surely if this is to distinguish mathematics from extraneous considerations regarding formal systems, he owes us an account. We should know what constructive reasoning—whether it is finitary or intuitionistic, for instance—is to count as nonextraneous mathematical reasoning.

Despite these limitations with Curry's presentation, I find the basic idea sufficiently attractive to attempt a more sympathetic reconstruction of it. Perhaps the following sketch will capture Curry's basic intuition. Ordinary mathematics with its somewhat dubious ontology and epistemology can be reduced to the theory of formal systems via the process of formalization. This is a definite ontological and epistemological gain, because the only commitments necessary in the theory of formal systems are to a potential infinity of (symbolic) objects and to elementary constructive reasoning about the same. Although various philosophical schools may debate about the status of these entities (e.g., whether they be physical, mental, or abstract) and the nature of this elementary reasoning (e.g., whether it be logical, intuitive a priori, or empirical), *no one contests their necessity for mathematics.* Thus, the *mathematical problem* of finding an adequate foundation for mathematics is solved. Mathematicians may proceed to study formal systems and *philosophers* may decide the metaphysical status of the minimal ontological and epistemological commitments required for mathematics.

This scheme fits very nicely with Curry's insistence that the "metaphysics" of formal systems is irrelevant for mathematics and explains his refusal to face ontological and epistemological issues. Nonetheless, it is open to at least two objections: one technical, the other philosophical.

The technical objection is that Curry's minimal rules for mathematical reasoning are insufficient for determining the truth-value of some *elementary propositions.* Let S be an elementary formal system in

[75]

which elementary number theory is formalizable and which meets the conditions of Gödel's second underivability theorem. Furthermore, let the assertion sign be the only predicate of S and let us suppose that S is consistent. Then "$\vdash 0 \neq$ " will be a true elementary proposition of S and "$\vdash 0 = 1$" will be a false elementary proposition of S. The former claim, however, can be established only by reasoning, which cannot be formalized in S. Thus, if S is sufficiently strong to formalize Curry's minimal reasoning, we can find an elementary proposition that can be known only by "extraneous" means. Obviously, neither Curry nor most mathematicians doubt that there are some elementary formal systems of this kind.

The more philosophical objection is one which we shall see again in connection with deductivism. It is this: in applied mathematics the validity of the "extraneous" considerations must be decided. For example, in using classical mathematics in physics, it must be determined whether there are domains in which the axioms of real analysis are satisfied. But these cannot be purely physical domains unless we countenance infinitely many physical objects. Thus, Curry's approach simply postpones the difficult questions about mathematical truth and foists them upon the physicist. Frege wrote, in connection with his plea that arithmetic be recognized as science:

> Otherwise it might happen that while this science handled its formulas simply as groups of figures without sense, a physicist wishing to apply them might assume quite without justification that they expressed a thought whose truth had been demonstrated. This would be—at best—to create the illusion of knowledge. [Frege 1903a:sec. 92]

Hilbert's Finitism

Hilbert's finitism antedates Curry's doctrines, but in some respects the former is a more sophisticated and thoroughly developed version of the latter. For Hilbert, too, formal systems and metamathematics play a major role, but mathematics is not merely the science of formal systems and classical mathematics is not an "extraneous" heuristic. In his concern to retain the scientific respectability of all of mathematics while meeting the constructivist's purges initiated by Brouwer, Hilbert attempted a theory of mathematical meaning and knowledge of considerable philosophical merit. His views and their attendant mathematical program had significant influence on research in the foundations of mathematics and were responsible for the development of the branch of mathematical logic known as proof theory. Because Hilbert's views underwent some evolution, it is useful to approach their exposition historically.

The first developments of Hilbert's philosophy of mathematics occurred around the turn of the century, when he wrote the famous *Foundations of Geometry*. This book contains the first set of complete and rigorous axioms for Euclidean geometry, together with proofs of their independence and consistency. Of course, the axiomatic approach to geometry traces to Euclid, so that there is nothing new in Hilbert's use of the axiomatic method. But Euclid's proofs were flawed by the presence of constructions that cannot be justified by his axioms. A rigorous axiomatization of the foundations of geometry was needed, and Hilbert provided a particularly simple one which retained the Euclidean spirit. Consistency and independence proofs were not new to Hilbert; they had been used by Klein in showing the consistency of non-Euclidean geometry. Indeed, Hilbert did not think he was making bold new contributions to mathematics in his book, which originated from a lecture course on the foundations of geometry given in the winter of 1899. Its success and influence were due more to the elegance of its approach and the systematic exposition of its results than to their dramatic originality (Reid:chap. VIII). Yet this book and Hilbert's influence persuaded many other mathematicians to adopt the axiomatic method in their researches.

Hilbert turned from geometry to the axiomatic approach to arithmetic, where a heated controversy raged over the "constructions" of Dedekind, Cantor, and Weierstrass of the real numbers via infinite processes. Here Hilbert employed his approach to geometry to make an entirely different attack on the problems of existence and truth in mathematics. Instead of seeking to justify the existence of mathematical entities by "constructing" them from some more basic entities, and to establish the truth of a theorem of a mathematical theory by reducing it to

[77]

the truth of some more elementary theory or statement. Hilbert suggested that we first axiomatize the theory in need of justification. Then, if we show that its axioms are consistent, we have done all that is necessary to show that they are true and that the entities of which they treat exist. To ask whether a real number with certain properties exists is simply to ask whether a certain existence theorem can be proved from the axioms of the real numbers. To ask whether a given statement of the theory of real numbers is true again amounts to asking whether that can be proved as a theorem from the axioms of real number theory. To ask in general whether there are real numbers or whether the theory of real numbers is true is, on Hilbert's view, simply to ask whether the axioms of that theory are consistent. (See Chapter Three for further details and references.)

This move had very liberating consequences for mathematics, because it allowed mathematicians to investigate any kind of mathematical theory without asking whether any "reality" corresponds to it. Until Hilbert's time the orthodox view of mathematics held that it had self-evident truths as its axioms. But the development of non-Euclidean geometry had rocked this orthodoxy, and the adoption of Hilbert's view removed the stigma attached to investigating non-Euclidean geometries or even nonstandard arithmetics and logics.

Despite the heuristic fruitfulness of this view, there are philosophical problems with it, especially as Hilbert originally stated it. How, for example, can consistency be sufficient for existence and truth? One can easily formulate a consistent theory of unicorns. That does not mean that unicorns exist or that statements about unicorns are true. Indeed, how can Euclidean and non-Euclidean geometry both be true? A closer study of Hilbert's writings at that time indicates that what he had in mind was something like this: A given mathematical theory studies a whole class of structures (i.e., those which statisfy its axioms). When working within a particular mathematical theory, the mathematician has no need to inquire into the nature of the structures which that mathematical theory treats. Other apparently incompatible axioms, such as is the case in geometry, simply deal with another class of structures. But what of the existence of the structures satisfying the axioms? This question threatens to send us again on a search for self-evident constructions.

Hilbert forstalled this by identifying consistency with the existence of the structures in question: "We recognize that [these axioms] never lead to any contradiction at all, and therefore speak of the thought-objects defined by means of them . . . as *consistent* notions or operations, or as *consistently existing* " (Hilbert 1904:134). Today, with our knowledge of the completeness theorem and of nonisomorphic models for various axiom systems, we can object that this identification tends to obscure distinctions and results that require difficult proofs. We may grant that the completeness theorem assures that every deductively consistent set of axioms is satisfied by *some* structure, so *in a sense* consistency is sufficient for existence and truth. On the other hand, the completeness theorem raises additional questions regarding Hilbert's view. The first is that all known proofs of the completeness theorem construct models from previously given entities, whether they be the natural numbers or the terms of the formal language in which the theory is formulated. So to be able to prove the completeness theorem, one must assume at the outset that a denumerable infinity of mathematical entities exists. Second, the principles used in the proof of the completeness theorem themselves need justification. It will not do to say that they can be collected into a consistent axiom system, because for the proof of the completeness theorem to be convincing, the principles used in its proof must be true. That *these* principles can be consistently axiomatized does not show that they are true, since the completeness theorem itself is supposed to bridge that gap. That point comes up again in the following way. To apply the completeness theorem, we have to show that the axioms are consistent, and if we show that the axioms are consistent, we have to use some principles which we cannot justify by simply saying that they are consistent. Thus, Hilbert must accept some mathematical entities as basic and some truths about them as unquestioned, or else give up the quest for consistency.

Frege raised an objection which Hilbert eventually managed to bypass successfully. Frege's objection was that there is no way to show that a set of axioms is consistent except by giving a model for them, so that Hilbert's criterion—even if it could be justified—would be useless. Giving a model for a set of axioms is to establish the existence of certain objects and to show that the axioms are true of them. Frege and Hilbert exchanged their views in correspondence on the question, and at that time Hilbert did not appear to have a ready answer. (See Chapter Three for references.) Frege's objection is not always applicable, however, since one can establish that one system is consistent provided that another is consistent by interpreting the former in the latter—as Hilbert had done. This was the only method Hilbert knew prior to 1904. Yet he also saw the need to stop the regress of relative consistency proofs with an absolute one, especially one for arithmetic, upon whose consistency that of geometry was predicated (Hilbert 1904:135). It is at this point that the force of Frege's objection begins to be felt.

In a paper of 1904, however, Hilbert started on the development of a response to both the two objections I gave earlier and to Frege's objection. He began by considering axiom systems as formal systems whose elements are discrete objects and can be subjected to mathematical investigation. This permitted the consistency questions to be formulated in terms of the derivability of formulas expressing a statement and its negation. This may seem a rather trivial point to those who have been trained in modern mathematical logic, but to get some appreciation of how momentous Hilbert's insight was, it should be noted that in discussing Thomae's formalism, Frege took inconsistency to be the presence of rules that both permit and prohibit symbol manipulation (Frege 1903a: secs. 113–120). Thus, Frege, who in 1904 had the object language and metalanguage distinction more firmly in mind than Hilbert did, narrowly missed formulating the consistency problem. It took some 20 years before it was satisfactorily stated. Hilbert also made a rather vague and imprecise attempt to set up an elementary formal system and give a consistency proof for it. Although the rigor of his procedure hardly approached Frege's, and certainly did not match what he himself would later use, Hilbert's paper contains many of the basic ideas of his proof theory.

A difficulty with this paper from the philosophical point of view is that Hilbert did not completely separate the reasoning to be used in proof theory from the axiomatic method. This comes through in his introduction of 1 as simply a "thought object" and his claim that the identity axioms *define* identity (Hilbert 1904:131–132). It appears, then, that Hilbert still was trying to develop mathematics as devoid of basic truth and ontology.

The 1904 paper was criticized by Poincaré and Brouwer. Both argued that Hilbert's suggestion that the foundations of mathematics can be secured by consistency proofs for various formal systems begged the question in an important sense. Because a consistency statement refers to *all* proofs in a system, mathematical induction will be needed in the metalanguage to establish it, and this is exactly the sort of principle that Hilbert sought to justify through his approach (Poincaré:464–467; Brouwer:71). This charge merits a response, since in the 1904 paper Hilbert did use an inductive argument and continued to do so in later proofs (Hilbert 1904:134). Brouwer leveled other charges against formalism, which are familiar from our discussion of the work of Thomae and Curry. These are: (1) the formalists do not explain why we are interested in certain systems and not others; (2) they must admit contradictory results as being mathematical; and (3) in setting up their systems they fall back on the assumption that there is a mind-independent world of mathematical objects to be described by the axioms.

Hilbert did not return to the foundations of mathematics until the 1920s, when he presented a series of papers that expounded and refined his program and responded to Brouwer and Poincaré. The crucial change in his position was his new goal of founding mathematics upon a basic set of mathematical entities and epistemologically evident truths concerning them:

> As we saw, abstract operations with general extensions of concepts and contents turned out to be inadequate and insecure. Rather something must already be given to our faculty of representation as a precondition for the use of logical inferences and the application of logical operations: certain extra-logically discrete objects which are intuitively present as immediately experienced prior to all thought. If logical inference is to be reliable then it must be possible to survey these objects in all their parts, and the fact that they occur, that they differ from one another, and that they follow each other, is immediately given intuitively together with the objects as something which cannot be reduced to anything else. In taking this stance the objects of number theory are for me—to the exact contrary of Frege and Dedekind—the signs themselves, whose form may be identified

by us with generality and reliability independently of the place, time, and the special conditions of the production of the signs as well as insignificant differences in its execution. Here lies the firm philosophical attitude which I take to be required for the foundations of pure mathematics—as well as for all scientific thinking, understanding and communication whatsoever: *in the beginning*—it could be said here—*is the sign.* [Hilbert 1922:163; cf. Hilbert 1925:376; 1927:465]

Hilbert's basic signs are the unary numerals, that is, the symbols "1," "11," "111," and so on, which, although meaningless in and of themselves, are the materials for meaningful mathematical study. Much like the procedure presented in discussing computation game formalism, Hilbert introduced ordinary decimal numbers to designate unary numerals. Thus, "2" designates "11," "3" designates "111," and so on. Meaningful statements can be formed using these designations; for instance, "2 < 3" asserts that the unary numeral designated by "3" extends beyond that designated by "2," "3 + 2 = 2 + 3" asserts that adjoining "11" to "111" yields the same result as adjoining "111" to "11." This approach may be extended to any other operation on, or predicate of, unary numerals whose results and truth-values, respectively, can be determined by finite and effective computation procedures. Thus, Hilbert certainly could—and probably did—recognize the meaningfulness of sentences in variable free primitive recursive arithmetic; for the truth-value of such sentences can be verified effectively by finite procedures operating upon the unary numerals. Hilbert took such a process to be epistemologically basic, independent of logic, and immediately evident to those who apply it (Hilbert 1922:163–164; 1925:377; 1927:470).

As our study of Thomae and Heine's attempts revealed, this approach is headed for difficulties when one attempts to extend it to the real numbers. Hilbert himself took Frege's critique to heart:

> Certainly we can proceed considerably into number theory with this intuitively meaningful type of procedure in the manner that we have sketched and used it. But of course, all of mathematics cannot be comprehended within this sort. The transition to the

[*82*]

standpoint of higher arithmetic and algebra, e.g., when we wish to make assertions about infinitely many numbers of functions, already denies these intuitive procedures. Since we cannot inscribe numerals or introduce abbreviations for infinitely many numbers; we would, as soon as we fail to consider this difficulty reach the absurdities which Frege correctly censured in his critical expositions of the traditional definitions of the irrational numbers. [Hilbert 1922:165]

Hilbert's intuitive procedures, however, impose even more stringent conditions than Frege recognized. Thus far we have seen that Hilbert can countenance variable free sentences of primitive recursive arithmetic and truth-functional compounds of these. He also countenanced *bounded* existential and universal quantifications such as "$(\exists x)$ ($x <$ 5 · x is prime)" and "(x) ($x < 10 \supset (\exists y)$ ($y < 2 \cdot x + y = 10$))," since these are equivalent to finite disjunctions and conjunctions, respectively, of the former type of sentence (Hilbert 1925:377–378). Hilbert found a place in his scheme for sentences of primitive recursive number theory with *free* variables. For these can be interpreted as schematic devices for asserting their instances. For example, the formula

(1) $(x + y = y + x)$

is meaningful as long as we interpret it as being a short form for

"$x + y = y + x$" yields a truth if numerals are
substituted for "x" and "y"

(Hilbert 1925:377; Hilbert used German letters as schematic letters). On the other hand, this device will not yield the negations of these general claims. In the example at hand, we cannot assert that (1) is false for some x and y, for we simply obtain "$-(x + y = y + x)$," which says that "$x + y = y + x$" is false for *all* substitutions. To obtain a real negation, we must use unbounded universal quantifiers, viz.,

$-(x)$ (y) $(x + y = y + x)$,

[83]

which takes us out of the domain of intuitively meaningful or, as Hilbert also said, *finitary* or *real* statements (Hilbert 1925:378). The reason for this is that an unbounded quantification cannot in general be verified via a finite procedure. Thus, *the domain of real statements is not closed under the application of logical operations.*

In the finitary domain, the laws of classical logic can be secured, since they can be independently verified. For example, if p is a closed finitary sentence, then either p or its negation can be finitely verified, so "$p \lor -p$" will be a finitary truth. Yet from finitary statements we can infer unbounded quantifications; for example, from "$1 + 3 = 4$" we can infer "$(\exists x)\ (x + 3 = 4)$." Thus, *the domain of finitary statements is not closed under application of the rules of inference of classical logic.* Furthermore, in the nonfinitary domain we have lost the "precondition" for the use of logic since, in general, we cannot independently verify its results in a finitary way.

In some instances a nonfinitary statement can be replaced by a finitary substitute. For example,

$$(x)\ (x \text{ is prime} \supset (\exists y)\ ((y > x) \cdot (y \text{ is prime})))$$

can be replaced by the finitary schema

$$x \text{ is prime} \supset (\exists y)\ ((y < x! + 1) \cdot (y \text{ is prime})).$$

Yet, as Hilbert was aware, this method certainly is inadequate for the greater part of classical mathematics. In other areas of mathematics closure conditions also have failed. For example, the rational numbers are not closed under the square-root operator; the real numbers are not closed under negative exponentiation; and in geometry two distinct lines do not always determine a point, although two distinct points always determine a line. Mathematics has confronted these irregularities by introducing new elements—ideal elements as Hilbert called them—such as the irrational numbers, imagery numbers, and points at infinity. This, in turn, has produced fruitful and elegant mathematical theories. Hilbert suggested that nonfinitary statements be viewed as ideal adjuncts to the finitary domain, with the purpose of guaranteeing the closure of the logical operators and rules of inference. The benefit to be received from them was a great one—the retention of classical mathematics (Hilbert 1925: 379; 1927:470).

One question that suggests itself at this point is: Why would Hilbert care about extending mathematics beyond the intuitively evident finitary domain? It is clear that Hilbert entertained no real doubts about classical analysis (Hilbert 1922:158; 1927:476) and even believed that set theory could be reclaimed from the paradoxes (Hilbert 1925:376–377). But can the introduction of nonfinitary ideal statements be justified only on the grounds that "we just do not want to renounce the use of the simple laws of Aristotelian logic" (Hilbert 1925:379) "since the construction of analysis is impossible without them" (Hilbert 1927:471)? This seems to beg the question, since the laws of classical logic in infinitary applications and classical analysis are the very things that cannot be directly justified from the finitary point of view. Hilbert was carried away here by his zeal to squelch Brouwer.

On the other hand, the introduction of ideal statements may be capable of indirect justification. Many ideal statements have real statements as logical consequences. For example, from the ideal "(x) (y) $(x + y = y + x)$" we may conclude the real "$1 + 2 = 2 + 1$" (Hilbert 1927:270). Without doubt, Hilbert did believe that the use of ideal statements to obtain further real ones furnished an indirect justification for classical higher mathematics, although in my opinion he did not sufficiently emphasize this:

> The science of mathematics is by no means exhausted by numerical equations and it cannot be reduced to these alone. One can claim, however, that it is an apparatus that must always yield correct results when applied to integers. But then we are obliged to investigate the structure of the apparatus sufficiently to make this fact apparent. And the only tool at our disposal in this investigation is the same as that used for the derivation of numerical equations in the construction of number theory itself, namely, a concern for concrete content, the finitist frame of mind. This scientific requirement can in fact be satisfied; that is, it is possible to obtain in a purely intuitive and finitary way, just

like the truths of number theory, those insights that guarantee the reliability of the mathematical apparatus. [Hilbert 1925:377]

This passage propounds an instrumentalist interpretation of nonfinitary mathematics. The ideal elements, although meaningless in and of themselves (Hilbert 1925:380), are elements of an instrument that facilitates our discovery of true finitary statements.

How essential is this instrument? The truth of each closed finitary statement can be determined by a finite calculation procedure, but obviously, calculations involving large finite integers can be so lengthy as to exceed human capacities. On the other hand, a short proof involving ideal statements could yield a quick solution. But is classical mathematics just a speed-up device which is theoretically dispensable? If it is, Hilbert's indirect justification is not very strong. The finitary schemata may provide the answer; for the use of finitary schemata brings indefinitely large numbers and infinitely many numbers within our finitary comprehension. Hilbert did not explicitly characterize finitary statements and finitary reasoning, but his discussion and examples support the choice of primitive recursive arithmetic (PRA) as the proper formalization of finitary mathematics. (In Van Heijenoort (p. 482) it is stated that Hilbert "came to recognize this arithmetic as the proper framework for the formalization of metamathematics.") To be definite, I shall assume that this is so; I believe that my subsequent claims about finitary reasoning hold for other formalizations as well. PRA allows us to express all finitary schemata and to prove many. If every *verifiable finitary schema*—that is, every one all of whose numerical instances are finitarily true—were a theorem of PRA, then it would appear that further mathematics would be dispensable both in theory and in practice. As will be developed below, however, there are important finitary schemata—ones Hilbert assumed to be verifiable—which are not provable in PRA. Ironically, although this seems to strengthen Hilbert's indirect justification for classical higher mathematics immensely, I will tie it to one of the most devastating failures of his research program.

In the last quotation from Hilbert we read that the mathematical apparatus must be justified through finitary means to guarantee its reliability. For him "reliability" meant that the use of ideal statements would not permit the derivation of false real statements, and he believed that this reliability could be secured by means of a finitary consistency proof for classical mathematics. The following passage is the fullest sketch of his reasoning:

> But even if one were not satisfied with consistency and had further scruples, he would at least have to acknowledge the significance of the consistency proof as a general method of obtaining finitary proofs from proofs of general theorems—say that of the character of Fermat's theorem. . . .
>
> Let us suppose, for example, that we had found, for Fermat's great theorem, a proof in which [nonfinitary reasoning] was used. We could then make a finitary proof out of it in the following way.
>
> Let us assume that numerals p, a, b, c, $(p > 2)$ satisfying Fermat's equation,
>
> $$a^p + b^p = c^p$$
>
> are given; then we could also obtain this equation as a provable formula by giving the form of a proof to the procedure by which we ascertain that the numerals $a^p + b^p$ and c^p coincide. On the other hand, according to our assumption we would have a proof of the formula
>
> $$(Z(x) \mathbin{\&} Z(y) \mathbin{\&} Z(z) \mathbin{\&} Z(w) \mathbin{\&} (p > 2)) \rightarrow (x^w + y^w \neq z^w),$$
>
> from which
>
> $$a^p + b^p \neq c^p$$
>
> is obtained by substitution and inference. Hence, both
>
> $$a^p + b^p = c^p \qquad \text{and} \qquad a^p + b^p \neq c^p$$

[*87*]

would be provable. But as the consistency proof shows in a finitary way, this cannot be the case. [Hilbert 1927:474; I have used "a," "b," "c," "p" in place of his German letters, and "x," "y," "z," "w" in place of his bound variables]

Let us note, first, that this passage contains a stronger claim than just that consistency is sufficient for the reliability of nonfinitary methods. It also maintains that—in some cases, at least—the finitary consistency proof will yield finitary proofs of results first established by nonfinitary means. Turning now to the analysis of Hilbert's argument, suppose that we have a finitary consistency proof for some formal system S, which is a nonfinitary extension of PRA. Suppose that the closed sentence "$(x)Fx$" has been proved in S and that "Fx" is a finitary schema. To prove it from the finitary point of view, we must show that it is verifiable. So suppose that we are given a numeral a and an instance "Fa." Now if "Fa" is not a finitary truth, a finite calculation will establish "$-Fa$." But this procedure can be formalized in PRA to give a proof in PRA of "$-Fa$," so "$-Fa$" will also be a theorem of S. But "Fa" is also a theorem of S, since "$(x)Fx$" is; and this contradicts the consistency of S. Thus, "Fa" must be finitarily true, no matter what numeral a is. As the reasoning used, given the finitary consistency proof, is finitary, we have a finitary proof for "Fx." (Cf. von Neumann 1931:52–53.) Since this argument also works for closed finitary sentences, we have shown that the finitary consistency of S is sufficient for its finitary soundness. More formally, if S is a system for which we have a finitary consistency proof and A is a finitary sentence or schema, then the schema

(1)   if x is a proof of A in S, then A

is verifiable. (Cf. Smorynski:Theorem 5.2.1.) We must take (1) in this form because "A is a theorem of S" is not a finitary sentence, since it involves implicit quantification over all proofs in S. For the same reason, we cannot use "$(\exists x)$ (x is a proof in S of A)" as the antecedent of (1), nor can we use the universal closure of (1).

In our discussion of why Hilbert took a finitary consistency proof to be an essential condition for the use of formal systems for higher mathematics (Hilbert 1925:383; 1927:471), we have so far assumed that *consistency* itself makes sense from the finitary point of view. Hilbert argued that this is possible because a consistency statement is about the formal system, not a statement within it; and that the formal system itself is a system of intuitively given symbols. So the epistemological status of statements about a formal system is the same as the epistemological status of statements about numbers (i.e., unary numerals). Just as there are finitely evident and meaningful statements about numbers, there are finitely evident and meaningful statements about formal systems (Hilbert 1925:380–383).

Thus, we must ask whether the statement that a given formal system is consistent can be interpreted as a finitary statement, that is, as a statement that does not involve unbounded quantification over the elements of a formal system. The answer is that it can be interpreted at least as a finitary *schema*. This requires some explanation. First, notice that to say that a given sequence of formulas is a proof of a given formula is to ascribe an effectively decidable relationship to these formulas, that is, one that can be decided through a finite process and/or direct inspection. Second, since the formal system in question contains classical logic, we know that it is consistent if and only if at least one of its sentences is not a theorem. So let us pick one that should not be a theorem, say, "$1 \neq 1$." Since the formal system contains PRA, "$1 = 1$" is already a theorem, and the formal system certainly will be inconsistent if its negation is a theorem. Thus, the consistency statements can be taken as the assertion that there is no proof in the formal system of "$1 \neq 1$." This statement is an unbounded universal quantification. But the assertion that a particular sequence is not a proof of "$1 \neq 1$" is a finitary statement. Therefore, the universal quantification over all these sequences can be interpreted as a finitary schema asserting that each individual proof sequence is not a proof of "$1 \neq 1$" (Hilbert 1925:383).

But to give a finitary sense to a consistency statement for a formal system and to give it a finitary proof are two entirely different matters. Hilbert thought that such proofs were possible, and during the 1920s some of his students showed that certain subsystems of elementary number theory have finitary consistency proofs. In 1931, however, Kurt

Gödel proved two theorems that completely disrupted Hilbert's program.

The first of these theorems states (roughly) that given any formal system S containing elementary number theory, one can effectively find a statement in the language of S such that neither it nor its negation are theorems of S, provided that S is consistent. These statements are statements about the natural numbers, and given the consistency of the system in question, metamathematical considerations show that some of these statements are true. Many people have concluded that this shows that there are number-theoretic truths that cannot be proved in any single formal system, and therefore Hilbert's goal of capturing all of number theory in a single formal system cannot be satisfied. But Gödel's first theorem is not devastating to Hilbert's program *as I have expounded it so far*. Hilbert did not identify truth with provability, and he did not claim that every mathematical statement—ideal as well as real—is either true or false. So he could accept with equanimity the existence of statements that are neither provable nor refutable within a given formal system, as long as these statements are ideal statements and are very far removed from ordinary mathematical consideration. (There is little reason to think that we cannot formalize in a single formal system the mathematics of practicing mathematicians—insofar as their work falls outside certain parts of mathematical logic.)

Gödel's first theorem *does* bear upon Hilbert's attempts to answer Brouwer. The latter disparaged the "game" aspects of formalism and inveighed against the unrestricted use of the law of the excluded middle which he took to be tantamount to the assumption of the solvability of all mathematical problems. In response, Hilbert wrote:

The demonstration that the assumption of the solvability of every mathematical problem is consistent falls entirely within the scope of our theory. [Hilbert 1925:384]

The formula game that Brouwer so deprecates has, besides its mathematical value, an important general philosophical significance. For this formula game is carried out according to certain rules, in which the *technique of our thinking* is expressed.

These rules form a closed system that can be discovered and definitively stated. [Hilbert 1927:475]

Admittedly, the phrasing here is too vague to establish definite conclusions about Hilbert's claim. Nonetheless, a plausible interpretation can be given. Since Hilbert regarded logic and mathematics as parallel disciplines, the principles of thought that he had in mind must have included mathematical as well as logical reasoning. Mathematical problems are to be solved via such reasoning. Thus, if it does form a "closed system" that can be captured in a formal system S, solving a mathematical problem will amount to proving a certain theorem of S, and the solvability of all mathematical problems will be equivalent to the formal completeness of S.

Gödel's first theorem shows that systems of this kind must be formally incomplete if they are consistent. In the passage quoted, Hilbert spoke only of establishing the *consistency* of the assumption that all mathematical problems are solvable. In this connection Kreisel has remarked that as an easy consequence of Gödel's second theorem, this (false) assumption may be added consistently to any consistent system to which the theorem applies, and that it is unlikely that Hilbert would be satisfied with this trivial proof (Kreisel:175). But regardless of whether Hilbert wanted to establish the assumption itself or merely its consistency, in the face of Gödel's incompleteness theorem, the formulation in terms of a particular formal system no longer makes sense. For the incompleteness theorems demonstrate that *there is no closed system of reasoning capable of solving all mathematical problems* (i.e., establishing all mathematical results).

Gödel's second theorem strikes at the core of Hilbert's program. This theorem states that the consistency statement for any formal system S, which itself contains elementary number theory, is not provable within S provided that S is consistent. This means that the consistency proof for elementary number theory, much less that for analysis and set theory, requires methods beyond the scope of elementary number theory. When this theorem was discovered, many foundational workers believed that finitary proof is a subspecies of elementary number-theoretic proof. Thus, many concluded that Gödel's second theorem dashed all hopes for a finitary consistency proof even for number theory. Later, Gentzen gave a proof of the consistency of elementary number theory which used only one principle that had not been previously recognized as finitary, namely, transfinite induction on ordinals less than the first epsilon number. Since some of Hilbert's followers found this principle intuitively convincing in the context of Gentzen's proof, the possibility was raised that the principles of finitary proof had not been fully characterized by subsystems of elementary number theory. (Nothing in Hilbert's writings supports this conjecture.)

The result of this hedging is that without a definitive characterization of finitist proof, there is little chance of establishing conclusively whether Hilbert's plan succeeds or fails. The formalization of mathematics is an open-ended activity involving a number of axiom systems. Although it is possible that the finitary consistency of some of these systems can be established by appealing to principles that belong to some other system, it seems unlikely that this is true of the more powerful systems used to formalize analysis or set theory. And this would definitely undermine Hilbert's claims. (For additional discussion of the technical aspects of Hilbert's program, see Kreisel. For a discussion of the philosophical aspects of consistency proofs, see Resnik (1974) and Detlefsen.)

Even without a final characterization of finitary reasoning, Gödel's theorems yield other information relevant to Hilbert's views which has not received sufficient notice. Suppose that S is a formal system that contains elementary number theory and any finitary reasoning not already present in elementary number theory. If S is consistent, then its finitary consistency schema

$$-(x \text{ is a proof in } S \text{ of } ``0 = 1")$$

is not one of its theorems, nor a fortiori is it finitarily provable. But this schema is provable in extensions of S. (It is trivially so if to S we add the new axiom ``$(x) - (x$ is a proof in S of '0 = 1'),'' but usually there are interesting extensions of S with this property, too.) On the other hand, if S is in fact consistent, then each particular finitary *sentence*

(1) $-(n \text{ is a proof of } ``0 = 1"),$

with n a numeral, can be established by an independent finitary proof. Thus, it appears that Hilbert could take comfort from this situation by arguing that higher mathematics (the extension of S in question) led us to new finitary truths, which were then independently verified by finitary methods. Indeed, the case at hand is very similar to the nonconstructive solution to Gordan's problem (Reid:chap. v), which first brought Hilbert into the mathematical limelight. Hilbert had this in mind in this passage urging the value of nonconstructive proofs:

> Now some time ago I stated a general theorem (1896) on algebraic forms that is a pure existence statement and by its very nature cannot be transformed into a statement involving constructibility. Purely by the use of this existence theorem I avoided the lengthy and unclear argumentation of Weierstrass and the highly complicated calculations of Dedekind, and in addition, I believe, my proof uncovers the inner reason for the validity of the assertions adumbrated by Gauss and formulated by Weierstrass and Dedekind. [Hilbert 1927:474]

Similarly, without a consistency proof for a consistent nonfinitary S, it will just happen that every attempt to give a finitary proof of a sentence of the form (1) will succeed. The "inner reason" for this, however, will probably take us even further—into the ideal domain of an extension of S capable of proving it consistent.

On the other hand, the apparent boon for Hilbert raises questions about his view of finitary schemata. It is clear that part of Hilbert's reason for accepting free-variable formulas of PRA into the finitary fold is that each can be construed as a schematic assertion of infinitely many finitary statements, that is, "as a hypothetical judgment that comes to assert something when a numeral is given" (Hilbert 1925:378). Yet this involves an implicit unbounded quantification over numerals in the metalanguage—no matter what numeral is substituted for x, "$x + 1 = 1 + x$" holds—and appears to take us beyond the finitary limits. Hilbert overlooked or ignored this, however, because he also believed that the assertion of a finitary schema would be grounded upon a finitary proof schema. This would be a proof structure involving the schematic letters in question, which would produce (with little or no modifications) a

[93]

concrete finitary proof of an instance of the schema when numerals replace the letters. This comes out quite clearly in his 1922 paper, wherein he follows the introduction of a schema with a proof schema (Hilbert 1922:164). These proof schemata presumably permit us to grasp what is on the surface an infinite process via finite means. On the other hand, unbounded quantifications are only ideal elements, because "one cannot after all try out all numbers" (Hilbert 1927:470).

Now, however, we see the need to distinguish at least two types of verifiable free-variable formulas: (1) those which are the conclusions of a finitary proof schema, and (2) those whose every instance is a finitary truth but which are not themselves conclusions of a finitary proof schema. Free-variable *theorems* of PRA are of type (1), while the finitary consistency schema for analysis is (probably) of type (2). *Hilbert cannot admit a formula of type (2) as meaningful* from the finitary point of view because it is false that "the contentual correctness of this communication can be proved by contentual influence" (Hilbert 1925:377). Hilbert uses this phrase while justifying the use of schemata. Here "contentual" is virtually synonymous with "finitary.")

Since a schema of type (2) is not finitarily meaningful, the class of provable and refutable formulas and sentences of PRA is exactly the class of finitary sentences and schemata—given our assumption that PRA captures finitary reasoning. Hilbert need not concern himself with statements outside this class, and thus Hilbert has no need for a mathematical apparatus beyond PRA! His indirect justification for classical mathematics has lost its punch. The bonus obtained from Gödel's work has turned out to be a Trojan horse!

Furthermore, the distinction between ideal and real statements can no longer be recognized as finitarily meaningful. There is no effective decision procedure for segregating the theorems of PRA from the nontheorems. Thus, there are no finitary means for determining whether a schema of PRA is finitary. Indeed, even without the argument of the last paragraph, Hilbert's distinction is in trouble, since there is no decision procedure for verifiable schemata either.

A final problem with Hilbert's schemata concerns his proof schemata and the role of induction in finitary reasoning. Brouwer and Poincaré took him to task, you will recall, because they believed that his method presupposed the very principle of induction which he sought to justify. Consider the following finitary derivation of

$$((x + 1) + 1 = (y + 1) + 1) \supset x = y$$

from

$$(x + 1 = y + 1) \supset x = y,$$

which does not involve induction:

> Let x and y be arbitrary numerals and assume that $(x + 1) + 1 = (y + 1) + 1$. Since $x + 1$ and $y + 1$ are numerals, too, our premise yields $x + 1 = y + 1$, and this yields, in turn, $x = y$.

Substitution of actual numerals for "x" and "y" in this schema will yield without any other changes a particular derivation of one finitary sentence from another. But Hilbert uses another type of proof, which cannot be fitted into this framework. The following passage contains an example of this kind of proof: his proof of "$x + y = y + x$" (again I will use "x" and "y" and "z" in place of Hilbert's German letters):

> Let—as we may assume—$y > x$, i.e., the numeral y extends beyond x: then y may be analyzed as of the form $x + z$ where z serves to communicate a number; we have then only to prove $x + x + z = x + z + x$, i.e., that $x + x + z$ is the same numeral as $x + z + x$. But this is the case so long as $x + z$ is the same sign as $z + x$, i.e., $x + z = z + x$. But here vis-à-vis the original communication $[x + y = y + x]$ at least one 1 has been disposed of by splitting it off, and this process of splitting off can only progress so far until the permuted summands are congruent with each other. Since each numeral x is composed from 1 and $+$ in the manner explained before, it can thus be decomposed again by splitting off and cutting out the individual signs.

> In pursuing number theory in this way there are no axioms

and thus no contradictions are possible. We just have here concrete signs as objects, we operate with these and make meaningful assertions about them. And in particular as far as the above proof of $x + y = y + x$ is concerned, this proof, as I would like to especially emphasize, is a procedure that rests exclusively upon the composition and decomposition of the numerals and is in essence different from that principle that, as the principle of mathematical induction, or inference from n to $n + 1$, plays such a prominent role in higher arithmetic. The latter principle is rather, as we will later recognize, a far reaching formula, a principle belonging to a higher level, which for its own sake requires proof and is capable of it. [Hilbert 1922:164]

One simply cannot get a proof of, say, "$6 + 8 = 8 + 6$" by substituting "6" and "8" for "x" and "y" in this argument. At best, one will obtain something which becomes a proof of this when it is adjoined to an extant proof of "$x + y = y + x$" for a smaller substitution for "x" or "y." For Hilbert to claim that this does not involve induction is to place too much faith in the naiveté of his audience! Indeed, in a subsequent paper Hilbert conceded that a form of induction was at work here:

Two distinct methods that proceed recursively come into play when the foundations of arithmetic are established, namely, on the one hand, the intuitive construction of the integer as numeral (to which there also corresponds, in reverse, the decomposition of any given numeral, or the decomposition of any concretely given array constructed just as a numeral is), that is, *contentual* induction, and, on the other hand, *formal* induction proper, which is based upon the induction axiom and through which alone the mathematical variable can begin to play its role in the formal system.

Poincaré arrives at his mistaken conviction by not distinguishing these two methods of induction, which are of entirely different kinds. [Hilbert 1927:472–473]

Contentual or finitary induction is a method of finitary reasoning which can be schematized as

$$F1, \ Fx \supset Fx + 1, \text{ therefore } Fx,$$

while formal induction is a method used *within* the formal system of arithmetic and takes the form of the axiom schema

$$(F1 \cdot (x) \ (Fx \supset Fx + 1)) \supset (x)Fx.$$

The crucial difference here is that finitary induction permits only the inference of a schema, whereas formal induction takes a leap into the transfinite with the unbounded universal quantifier in its conclusion. Furthermore, formal induction is, indeed, a stronger principle; there are inductive proofs that depend upon the use of the quantified conclusion and would fail if the free variable form were substituted for it. Here "the mathematical variable," as opposed to a schematic letter, "can begin to play its role." (See Steiner (pp. 139–154) for a detailed analysis.) Hilbert has a start—at least—on an answer to Brouwer's objection.

To be convincing, contentual induction must be seen to be evident from the finitary point of view. Hilbert's references to the decomposition of numerals suggest that he had in mind a justification for induction that might run as follows. Suppose that $F1$ and $Fx \supset Fx + 1$ have been established. Then Fx can be established by reflecting upon the fact that x itself is identical to $1 + 1 + 1 \cdots + 1$ for some definite finite number of terms. For $F1 + 1 + 1 + \cdots + 1$ holds provided that a similar statement holds with one less term, which in turn holds provided that a similar statement holds with still another one less term, and so on, until we arrive at $F1$. But this does hold, so Fx must hold for our particular x.

It has been objected repeatedly that this sort of justification is circular, since it involves a form of induction itself. For we need to be assured that this procedure can be successfully applied to *every numeral,* and that requires induction (Kitcher 1976:112–113). Hilbert might reply that the success of our procedure is grounded in the surveyability of finitary objects, and that here we are drawing a finitary conclusion based upon the finitary insight that each numeral is surveyable. But the surveyability of every numeral is itself one of the most dubious of Hilbert's epistemological premises.

This prompts us to ask about the more philosophical aspects of Hilbert's theory. What about his ontology of unary numerals? His emphasis upon their concreteness leads one to believe that they are physical inscriptions. Yet this is incompatible with his practice and some of his own remarks. For example, in promulgating his insight that "in the beginning . . . is the sign," he adds a footnote:

> In this sense I call signs of the same form "the same sign" for short. [Hilbert 1922:163]

This helps us make sense of Hilbert's claims that the signs on both sides of a true identity are the same and that in an inequality one sign is part of or extends beyond another. Moreover, Hilbert could do all this without committing himself to nonphysical symbols because his talk of a sign could be interpreted as talk of an inscription of a certain form.

On the other hand, Hilbert's ontology is at least potentially infinite and Hilbert himself argued that potential infinities do not exist in nature (Hilbert 1925:371-372). Nevertheless, he placed no upper bounds on the length of a numeral and therefore no upper bounds on the totality of the numerals. Given any numeral, one can construct an additional numeral simply by appending a "1" to it. Of course, certain parts of what Hilbert accepted as meaningful mathematics would remain true if there were an upper bound to the length of numerals; some statements of equality or inequality still would be possible. But one would have to give a very different characterization of addition and multiplication, because on his view these operations produce numerals that are longer than those to which they are applied. It is possible to handle this problem by using modular addition rather than ordinary addition. But then it would be difficult to explain how real arithmetic fits in with ideal arithmetic, for ordinary arithmetic cannot be construed as a simple, smooth extension of modular arithmetics. In any case, this certainly was not Hilbert's intention, as can be seen by the characterizations he gives of finitary arithmetic.

In view of this, it becomes very difficult to interpret Hilbert's claim that arithmetic deals with concrete objects. Since Hilbert spoke favorably of Kant (Hilbert 1925:376), it is possible that Hilbert thought of written numerals as Kantian constructions in intuition. If so, Hilbert's view is at least somewhat more plausible and clear than Kant's, because he does not extend it, as Kant did, to algebra. (But see Kitcher 1976.)

Whether Hilbert's epistemology and ontology is basically Kantian, it is open to a question which Frege raised with respect to Kant (Frege 1884:sec. 5). How do we see by direct inspection that a numeral of length 1,000,003 is shorter than a numeral of length 1,000,004? (If you think that we could see this directly, just assume that our numerals are sufficiently large that they cannot be taken in one glance but must be scanned.) If we have to scan the numerals to inspect them, we have to go through some kind of process to compare their lengths and to guarantee that it is not just some accidental feature that made the lengths come out the way they did. For example, it will not do to write one of the numerals above the other in a random way and then scan them to see which one stops first, because the digits of the top numeral might be closer together than those of the bottom numeral. It is clear that we have to use a one-to-one correlation in some form—for example, by making sure that we write the digits of the numerals directly above each other—so that we can avoid shrinkage or expansion problems. More complex situations will draw upon other properties of one-to-one correlation, such as its transitivity. This means that our basic arithmetical observations will be dependent upon at least a rudimentary theory. But this surely conflicts with Hilbert's claim that "mathematics is a presuppositionless science" (Hilbert 1927:479).

The only way out of this is for Hilbert to make more use of his criterion of surveyability; for he said that we must deal not only with concrete extralogical objects, but also only with configurations of them which are surveyable. His response might take two directions. He might reaffirm his principle that all numerals are surveyable and thereby dismiss my counterexamples as unreal. The only advantage to this is that it would also support his view of induction. Or he might concede that an upper bound upon the construction of numerals exists. Not only would this force him back to modular arithmetic, but it would press him to draw the line between a surveyable and nonsurveyable number. And the boundary might change with technology or evolution. We might be able to take in longer numerals in one glance by miniaturizing inscriptions of them and then using a magnification device. Our brains might develop through training or evolution to the point at which we could see numerals ''in all their parts'' which we could not see before. On the other hand, if we depend upon technology to survey longer numerals, how would we guarantee the reliability of our observations? Certainly not through ideal number theory!

Clearly, Hilbert would be in a better position if he took the stance that all numerals are surveyable and then gave *that* a less dogmatic defense. This would, I think, lead him to a reinterpretation of the finitary insights in which the axioms of PRA would play a more important role than they do in his original statement. More specifically, he might argue that our operations with short numerals reveal patterns that we ''see'' must hold for larger numerals and permit us to ''survey'' them as well. For example, in rearranging the terms of small sums we observe a pattern which we express by asserting the general associative and commutative laws of addition for all numerals. Similarly, by operating with $F1$ and $Fx \supset Fx + 1$ for small choices of x, we observe a pattern which shows us that Fx must hold for larger x as well. Finitary statements about small numerals would be directly evident, as would be the patterns recognizable through their aid. Finitary statements involving large numerals would then be justified by applying these patterns or general laws. The use of logic presupposed here could be justified, moreover, by appealing to the fact that our conclusions can be verified independently, at least in theory. Hilbert's ontological problems could be straightened out by recognizing abstract symbol types whose properties are known through observing the patterns to which their inscribed instances conform. In addition, it could be argued even that this new approach yields a priori knowledge, since pattern recognition allows us to abstract from the peculiarities of pattern instances. What we learn about a pattern itself is thus made independent of any particular experiences of its instances, although some experience is a precondition for the acquisition of this knowledge. Pattern recognition also seems to have the immediacy needed to ground the self-evidence that Hilbert sought. Furthermore, although numerals would now be abstract entities, finitary knowledge of them is so closely tied to operations with inscriptions that they are at least epistemologically concrete.

This reinterpretation of Hilbert does justice to his pronouncements. We can give a sense to the concreteness of his numerals and their surveyability while admitting their potential infinity. It also fits well with his views of finitary schemata. Furthermore, although much is still unknown about pattern recognition and the epistemological status of claims based upon it, it is no mysterious Kantian construction or Platonic intellectual intuition. It is a phenomenon that can be observed in everyday life—an important element in our knowledge of language and music—and no mere philosophical speculation but a subject of scientific study.

Hilbert's distinction between real and ideal statements foundered on Gödel's results. My reinterpretation of him puts less emphasis upon the direct verification of all finitary sentences and, accordingly, less emphasis upon the notion of verifiable schemata. Nevertheless, difficulties concerning the distinction between finitary and nonfinitary schemata continue to arise. How, for example, can we determine effectively whether a schema does express a pattern? We cannot if we equate this with provability in PRA. And what other choices do we have? And how are we to guard against the possibility that other essentially nonfinitary patterns might be recognized? It is plausible to view all claims to have a clear "intuition" of a mathematical structure as cases of pattern recognition. There are plenty of mathematicians who claim to have firm intuitions concerning such nonfinitary structures as the continuum and the iterative hierarchy of sets. Every increase in our knowledge of these structures makes it more difficult to defend an epistemological distinction in kind between the statements of set theory or analysis and those of number theory. It seems much more plausible to be more holistic and regard some statements of number theory *and* analysis *and* set theory as more evident than other statements of any of these disciplines. Ironically, Hilbert *might have* hinted at a similar view:

This formula game enables us to express *the entire thought-content of the science of mathematics* in a uniform manner and develop it in such a way that, at the same time, the interconnections between the individual propositions and facts become clear. To *make it a universal requirement that each individual formula then be interpretable by itself is by no means reasonable;* on the contrary a theory by its very nature is such that we do not need to fall back upon intuition or meaning in the midst of some argument. What the physicist demands precisely of a theory is that particular propositions be derived from laws of nature or hypotheses solely by inferences, hence on the basis of a pure formula game, without extraneous considerations being adduced. Only certain combinations and consequences of the physical laws can be checked by experiment—just as in my proof theory *only the real propositions are directly capable of verification.* [Hilbert 1927:475; my emphasis]

I say "might have" because this passage can be interpreted as nonholistic by taking the "thought-content of mathematics" to be just the finitary domain and the rest of it as a systematizing device.

The two readings to which Hilbert's passage gives rise are, I think, symptomatic of an underlying vacillation on his part concerning classical mathematics. On the one hand, it seems to him to be in need of a justification—one without which it must be discarded. On the other hand, he never seriously questions its claims to remain in the mathematical fold. For to raise this question would be to face the Frege-Brouwer question: If classical mathematics is *just* an instrument, what about the claims of nonclassical mathematics?

It will be useful to conclude this discussion of Hilbert with an overview of the previous discussion of other branches of formalism as well. Thomae's computation game formalism met its technical demise in the theory of real numbers. Frege demonstrated this, and Hilbert accepted his critique, although he tried to revamp the more elementary formalist insights. Derivation game formalism was made possible by Frege's codification and formalization of the principles of logic requisite to mathematics. Technically, it is an advance over Thomae's view, but philosophically they are on a par, since both fall prey to Frege's protest that mathematics is a science and must be so viewed to explain its applicability and nonarbitrariness. The latter part of this objection came to haunt Hilbert when Brouwer charged that he had no means for explaining his choice of ideal axioms unless, of course, he tacitly presupposed their meaningfulness. In the last point, Brouwer echoed Frege's charge that the formalists tacitly used meaningful arithmetic.

Curry tried to make a serious science of mathematics by giving it a subject matter. The trouble is, he tried to do this while purging it of ontology and epistemology—and that did not come off. We construed Curry's view so that it shows how mathematics in its syntactic aspects can be reduced to an uncontested ontological and epistemological minimum. In the process we approached Hilbert's views. But Curry did not answer the Brouwer-Frege objection fully either, since he gave no explanation of why mathematicians take some formal systems so seriously while disregarding others.

Hilbert attempted the latter and also promised to make up for Curry's faults. Mathematics was given an ontology and epistemology—his concrete extralogical objects and finitary reasoning. Part of mathematics—the ideal part—was treated along derivation game lines. But its claim to be a serious science was advanced also, on instrumentalist grounds, and its reliability as an instrument was to be secured via a finitary consistency proof. It was a bold and ingenious idea, even if it did not fully answer either Brouwer or Frege. For the former could complain that Hilbert did not demonstrate the uniqueness of his ideal realm, and the latter could demand to know how one could possibly infer a meaningful real statement from an ideal one. But in comparison with the power of Hilbert's conception, these objections seem minor and idiosyncratic. Hilbert's view is not susceptible, by the way, to the objections I raised earlier to general instrumental formalism, since he does recognize some meaningful mathematical truths and uses them to establish the properties of his instrument. His formalism is a tailor-made response to my objection, or in truth my objection is tailor-made to motivate his formalism!

Hilbert's view fails both technically and philosophically. The technical failures are due to Gödel's and related results which eradicate the possibility of finitary consistency proofs for significant branches of mathematics and discredit the distinction between ideal and real statements. From the philosophical point of view, there are difficulties with Hilbert's "concrete objects," which fall into the Fregean trap for potential infinities. Then there is also the Fregean problem of the self-evidence of finitary statements about large numbers, as well as the further epistemological problem over induction and surveyability.

This can be repaired in part while still retaining much of the Hilbertain spirit by turning to abstract symbols and pattern recognition. There is a gain in clarity here but a loss of finitary vision, which is incapable of a post-Gödel recovery anyhow. In opening the door to abstract entities, we may also have let in elements of the ideal domain; and we seem driven toward a mathematical holism. It is a holism with degrees of evidentness, one in which both ideal and real statements could be evident to the highest degree. Of course, *every* formula now being meaningful, we can hardly call this formalism.

Deductivism

Deductivism approaches the problem of mathematical existence, meaning, and truth by concentrating on the centrality of proof to the methodology of mathematics. Owing to the influence of Hilbert, Frege, and other pioneers in axiomatics, the current official position in mathematics is that a result of a given branch of mathematics cannot be considered as established until it has been deduced from the axioms characterizing that branch. The deductivist extracts from this position a very plausible view of the form and meaning of a mathematical statement: that a statement S of a mathematical theory T is implicitly of the form "S is deducible from $AX(T)$," where "$AX(T)$" is a characterization of the axioms of T. This immediately permits the deductivist to reduce the de jure epistemology of mathematics to that of deduction—which is usually taken to involve only first-order logic—thereby apparently dispensing with the need for Kantian or Platonic intuition or empirical induction. Furthermore, because a mathematical deduction from axioms can be carried out without any presuppositions concerning the truth of the axioms or their ontology, the deductivist can adopt a stance of agnosticism concerning the existence of mathematical objects and truths. This standpoint can be tidied up by construing the primitive nonlogical terms of mathematics as variables or schematic letters. Deductivism thus takes the viewpoint assumed in abstract algebra and generalizes it to account for all of mathematics.

Deductivism is closely related to formalism. Historically, prominent advocates of formalism have also leaned toward deductivism. As we have seen and will see in more detail shortly, Hilbert went through a deductivist period. Even Curry's agnosticism with respect to the ontology of formal systems can be interpreted along deductivist lines as long as mathematical induction is permitted as a method of deduction. For what could make better sense of his claim that mathematics is concerned primarily with laying down inductive definitions and exploring their consequences? Philosophically, deductivism is close to formalism in seeking to keep meaning and reference in mathematics to a minimum. Traditional meaning is denied to at least the nonlogical primitives of mathematics and, depending upon whether these are in turn taken as schematic letters or variables, entire theories can be disinterpreted. Deductivism also parallels formalism in permitting the choice of a set of mathematical axioms to be in principle an arbitrary choice. Unlike formalism, however, deductivism does not place particular emphasis on the formulas of mathematics per se. It takes mathematics to be neither a game with formulas nor the study of them.

Frege had an important role in clarifying the foundations of deductivism. This came about through his involvement in a controversy with Hilbert concerning the latter's *Foundations of Geometry*. We have mentioned this book and its importance both to the development of Hilbert's thought and to approaches to the foundations of mathematics. The study of the Frege-Hilbert controversy should furnish even deeper insights into Hilbert's doctrines. Surprisingly, although both parties came away from this discussion with a mathematically adequate formulation of deductivism, neither adopted it. Frege fell back on the more traditional view of mathematics as a science with self-evident and true axioms, while Hilbert moved on to finitism as his ultimate foundational proposal. (The axiomatic method still played an important part in Hilbert's views, but on the epistemological side he even went so far as to argue, you will recall, that logic needs a mathematical investigation to secure its use in mathematics.)

My plan in this chapter is to present the Frege-Hilbert controversy first and then to turn to a contemporary defense of deductivism given by Hilary Putnam. I shall close with a general discussion of deductivism.

The Frege-Hilbert Controversy

This controversy began when Frege's examination of Hilbert's *Grundlagen der Geometrie* (1900) left him so profoundly dissatisfied

[*106*]

with the book that he initiated a correspondence with Hilbert to express his doubts and to seek clarification.

After Hilbert discontinued this, Frege published two series of articles—each entitled "Über die Grundlagen der Geometrie"—in which he analyzed Hilbert's methods and presented his own theories concerning axioms and their consistency and independence. These papers show that Frege apparently had a clearer grasp of Hilbert's method than Hilbert himself did; indeed, he actually anticipated some of Hilbert's ideas. Much of the confusion concerning "implicit definitions" can be traced to Hilbert, while Frege's essays contain excellent accounts of the roles of axioms and definitions in mathematics.

The Nature of the Axiomatic Method

Frege and Hilbert agreed upon the general formulation of the axiomatic method by insisting that in the deductive development of a theory, all assumptions that are to be used without proof must be stated in advance. They also deplored their contemporaries' practice of adding new axioms in midtheory without bothering to determine whether this altered the theory in question. Through this unrigorous "genetic method," the theory of the natural numbers, for instance, evolves through the addition of axioms into the theory of the integers, and thence to rational and real number theory (Frege 1941:412, 414; 1893:VI). Finally, both accepted the (then) revolutionary ideal of complete formalization, which permits proofs to be checked mechanically without reference to the meaning of any of the symbols used. (This ideal, although already explicit in Frege, begins to emerge in Hilbert's letter to Frege. Of course, it later became a cardinal point of Hilbert's finitistic program.)

Derivability questions led the two men to their conceptions of the axiomatic method. Hilbert believed that his axiomatic method created the possibility for a rigorous investigation of the independence of the parallel postulate (Frege 1941:410). Frege, on the other hand, emphasized that his formal system was necessary to demonstrate the *dependence* of arithmetic upon logic (Frege 1884:sec. 90; 1893:VII).

Although Frege and Hilbert agreed upon many general points, there were deep differences in their philosophical attitudes toward the axiomatic method. Frege's enterprise started with a philosophical analysis of the concept of number, which in turn led him to use formalization to establish the correctness of his analysis. Hilbert, by contrast, appears to have been prompted by the mathematical problem of discovering a consistent and independent set of axioms for geometry. His philosophical views probably developed after he solved this problem, and were tied to its solution. These differences in the motivation of the two men will not answer all the questions that the controversy raises, but we will see that several of their disputes about methodology can be traced to fundamental philosophical differences.

Surely at that time one of the most perplexing problems of the axiomatic method concerned the status of the axioms themselves. It was quickly seen that they are not (and cannot be) proved (nontrivially) within the theory which they axiomatize. But then what justification do they receive? And what force do proofs based upon them carry?

Frege's answer was traditional and conservative. To avoid circularity in proofs we must, of course, take some truths as unproved. However, the source of these truths must be sought outside the theory being axiomatized. In the case of Euclidean geometry, the axioms are based upon a (Kantian) intuition (Frege 1941:409; 1912:337). Surprisingly, Frege never developed an epistemology for logic, and explicitly stated that logic could not give a noncircular answer to this question (Frege 1893:xvii). Moreover, in many of his writings Frege seems to suggest that axioms must not only be true but must also possess a special property, which might be called "self-evidence" or "obviousness." This implies that one cannot select just any subset of the theorems of a theory as its axioms, even if this set is sufficient for generating the others. (The details of Frege's position will be discussed in Chapter Five.)

Hilbert's philosophically and mathematically creative answer to the question of the status of the axioms must have been generated from the tension created in mathematical thinking by the discovery of non-Euclidean geometries. He claimed that we need only demand that our axioms be consistent. He shifted the emphasis in mathematics from questions of truth to questions of deductive relationships. He removed the stigma of investigating axioms that do not describe any known "reality" and opened the way to the creation of new mathematical theories simply by laying down new axioms. In philosophy this lent credence and respectability to conventionalism. (This is not to say that Hilbert was at that time, or ever became, a conventionalist.)

Definitions were also a cloudy subject at the time of the controversy. For it was recognized that just as circularity must be avoided with respect to truth (provability), it must also be avoided with respect to meaning (definability). Frege had a much more explicit theory of definitions than did Hilbert. Since he claimed that the laws of arithmetic are definitional transforms of laws of logic, he strongly favored the view that definitions simply introduce abbreviations (Frege 1879b:55; 1893:45; 1903b:263). He also presented precise formal rules of definition, which ensured that his definitions would satisfy the requirements of eliminability and noncreativity (Frege 1893:51-52). (Frege did not explicitly formulate these criteria; they came later and were due to Lesniewski. His own use of the term "creative definition" refers to mislabeling of an existence postulate as a definition. See, for example, Frege 1903a:secs. 139-147.)

With Hilbert, Frege realized that all definitions must rest ultimately upon expressions that remain undefined and for which no definition (in the system in question) is possible. Also, with Hilbert, he rejected as definitions such traditional examples as "a point is that which has no parts" or "a line is that which has no breadth," which were used to "define" the primitives of Euclidean geometry.

How, then, do primitive terms receive any meaning? Frege's view is that they must be elucidated, that is, explained but not formally defined. Here is Frege's own exposition:

[Elucidation] is used to enable scientists to understand each other and to communicate science. It is to be counted as part of the propaedeutic. It has no place in the system of science; no inference is based upon it. Anyone who wishes to do research for and by himself has no need for it. The goal of elucidation is a practical one, and if it is attained one must be satisfied with it. In doing this we must count on good will, halfway meetings of the understandings and hints: since without the use of metaphors we may never get anywhere. But it must be required of the elucidator that he knows definitely what he means, that he always agrees with himself, and that he is always ready, if the possibility of a misunderstanding arises (even when he is met with good will) to make his elucidation more complete and perfected.

Since a collective enterprise in science is not possible without
the mutual understanding of the researchers, we must trust that
such an understanding can be reached by elucidation, although
theoretically the contrary is not excluded. [Frege 1906a:288]

Frege is not employing a very precise notion here, since in the appropriate context almost anything can count as an elucidation. One may
consist in metaphorical half-truths—as in the planetary model of the
electron—or it may involve even the informal presentation of some
truths of the theory whose primitives are being elucidated. This technique is common in the case of very abstract theories, such as set
theories, since their terms have to be learned mainly in context (Quine
1960:13–17). Frege differentiated clearly between definitions and
elucidations. Only the former are to be reckoned as part of an axiomatic
theory and must as a consequence satisfy rigid forms. While thus rejecting Euclid's "a point is that which has no parts" as a definition, Frege
could have accepted it as an elucidation.

Hilbert's approach to the problem of primitive terms was as radical
and innovative as was his treatment of axioms. His view was that the
axioms (partially) define the primitive terms, since "obviously every
theory is only a framework or schema of concepts together with their
necessary relationship with each other and the primitive elements can be
taken arbitrarily" (Frege 1941:412). This was a very creative step, for it
not only gave an additional impetus to the ideal of formalization (via
disinterpretation) but also led to the development of modern postulational theory.

Nonetheless, Hilbert's position, as it stands, is in need of clarification. How can *axioms* be definitions? And what sorts of definitions are
they if they fail to determine uniquely the terms they introduce? These
questions form the basis for Frege's critique.

Axioms vs. Definitions

When Frege first read the lecture notes that became Hilbert's *Grundlagen,* he based his interpretation solely upon the text before him.
Hilbert's claim that his axioms expressed "fundamental facts of our
intuition" immediately struck a responsive chord (Hilbert 1899:3).
However, later Frege read that the axioms were also supposed to *define*

"the idea expressed by the word 'between' and that expressed by the word 'congruence'" (Hilbert 1899:5). To make matters more puzzling, he found paragraphs entitled "Definitions" in which terms such as "broken line" and "triangle" were given standard explicit definitions. Frege, perhaps the most rigorous mathematician of his day, was so overwhelmed by these passages that he wrote to Heinrich Liebmann that the entire work had "failed and in any case is only applicable with very much criticism" (Frege 1940:397). He also wrote a somewhat patronizing letter to Hilbert which, although primarily a lecture on the axiomatic method, accused him of confusing axioms with definitions and failing to make proper use of elucidations (Frege 1941). Given the data before him, Frege's criticism was completely justified.

As one of the foremost apologists for rigor (Hilbert 1900a) Hilbert must have been irked by this letter. But he exercised patience in his reply and gave Frege two hints for understanding his conception of axioms. First, he stated that his *total* system of axioms defined his primitives, with each axiom contributing a characteristic of the terms it defined. Second, he stated that his axioms did not uniquely determine any single concept—that *this* was their great mathematical advantage. The axioms actually determine many systems of concepts, each of which will satisfy the axioms under an appropriate one-to-one mapping between the primitive terms and the concepts of the system in question. Thus, in this letter Hilbert stated the basic idea of postulational theory. Frege was also referred to Hilbert's short article "Über den Zahlbegriff." While Hilbert did not explain his axiomatic method in any detail in this paper, he did lay down axioms and state that the real numbers are not, say, convergent sequences of rational numbers or Dedekind cuts, but simply a system of things whose mutual relations are given by the axioms. (Although this is very close to the model theoretic approach of postulation theory, it definitely is not completely so. For Hilbert continued by stating that any statement that can be *derived* through a finite number of logical inferences from the axioms also holds for the real numbers. He did not employ the notion of a semantic consequence (Hilbert 1900b).)

Frege read this article and, in his second letter to Hilbert, wrote that he now understood that Hilbert was attempting to free geometry from intuition and make it a matter of pure logic. This was accomplished by construing the so-called axioms not as unproved truths but as (implicit) conditions on every geometrical theorem. Thus, if A represents the conjunction of the axioms, then a theorem T in Hilbert's system should be expressed fully as $A \supset T$. The question of the truth of T is thus replaced by the question of the logical provability of $A \supset T$. Furthermore, the primitive terms of geometry can be viewed now as (predicate) variables, so that Hilbert's geometry becomes part of pure logic, with the generality of its application being an immediate consequence.

Frege construed the conditional $A \supset T$ as a second-order universally quantified conditional. The primitive terms "point," "line," and so on, are replaced by variables, and universal quantifiers binding these are understood as implicitly prefixed to the whole conditional. The theorems of the various "models" for the "axioms" are obtained then from these second-order quantifications by applying universal instantiation and *modus ponens*. By contrast, Hilbert seems to have viewed the primitives "point," "line," and so on, as schematic letters rather than implicitly bound variables, thus paving the way for a model theoretic approach. Of course, this was not explicit in Hilbert's writings (Frege 1941:413).

Nonetheless, the problem of the status of Hilbert's axioms as definitions still remains. As Frege remarked (Frege 1941:415–416), the major difficulty here is that they fail to determine a unique model, thereby leaving the meanings of "point," "line," and so on, undetermined. Another problem is the existential character of some of the axioms. For instance, it follows from them that at least two points exist on every line. Are we to conclude that these points exist by definition (Frege 1941:416)? Fortunately, this problem furnished the clue for Frege's solution to it. Frege conceived of quantifiers as higher-order predicates (Frege 1893:36–38). The terms "point," "line," "between," and so on, are first-order predicates. Hilbert's axioms contain both these first-order predicates and second-order predicates, namely, quantifiers. But the first-order predicates "point," "line," and so on, may not only be viewed as variables; they can also be construed as marking argument places in the second-order predicates of Hilbert's axioms. Thus, the conjunction of these axioms can be construed not as defining several first-order predicates but as defining a single second-order relational predicate (Frege 1941:416; 1903b). Thus, suppose that in the conjunction of the axioms, the terms "point," "line," "plane," "between," and "congruence" are replaced by the predicate letters "F," "G," "H," "I," and "J." Let the resulting expression be

$A(F, G, H, I, J)$.

Then (in non-Fregean terms) Hilbert's axioms can be interpreted as defining the notion of an Euclidean structure—that is, an ordered quintuple $\langle F, G, H, I, J \rangle$ such that F, G, and H are classes, I is a ternary relation, J is a binary relation and $A(F, G. H, I, J)$ holds. In a similar manner, a group can be defined as an ordered pair $\langle D, f \rangle$, where D is a domain and f an operation defined on D such that D and f satisfy the axioms of group theory.

Although rudiments of these two interpretations of "implicit definitions" appear in Frege's letter to Hilbert, he did not fully explain them there. Perhaps because of this and other fundamental disagreements, Hilbert broke off the correspondence. Frege subsequently published two series of articles in which his conceptions were explained completely (Frege 1903b; 1906a). In Hilbert and Bernays's *Grundlagen der Mathematik,* Volume I (1934), there is a description of the axiomatic method which (except for avoiding higher-order logics) agrees with Frege's account of Hilbert's reduction of geometry to pure logic. On the other hand, I have been unable to locate a statement by Hilbert that his axioms define structures or second-order relations. However, Bernays does acknowledge the correctness of Frege's account on this point (Bernays).

It is hard to believe that Hilbert did not have a clearer understanding of his own work than the documents indicate. The ideas surely were familiar to people working in algebra, and nearly a century before Gergonne introduced the term "implicit definition" and the idea of freely postulated axioms (Freudenthal). Perhaps Frege's apparently superior understanding of these matters is due to his possession of a sophisticated terminology with which to express himself. In any case, Frege was the first to give written evidence of an adequate grasp of the use of implicit definitions.

Despite this, Frege remained wholly unappreciative of the potential of Hilbert's work for mathematics in general and credited it with no relevance for traditional geometry. Indeed, he even dismissed Hilbert's independence results because they treated only defining conditions and not genuine axioms. To us this is shocking, since Hilbert's reduction of axiomatic theories to pure logic is as basic to mathematical logic as Frege's reduction of arithmetic to set theory. It is natural to ask why Frege was so blind to this.

Of course, part of the answer can be found in Frege's ontological attitudes. Euclidean geometry for him truly described spatial intuition, and Hilbert purposely avoided this entanglement. But leaving geometry aside, to Frege, Hilbert's conception of the axiomatic method completely bypassed the epistemological status of mathematical axioms. Indeed, as far as the theory of implicit definitions is concerned, there could be both empirical and a priori Euclidean structures or even systems of real numbers.

In this context it is particularly interesting to note that Frege anticipated Hilbert's reduction in 1884. He remarked in passing that one could reduce arithmetic to logic in the way in which geometry can be: by showing that the theorems can be derived from the axioms by logic alone (Frege 1884:23–24). But he did not pursue this because it still did not determine whether the axioms themselves can be reduced to logic. This passage provides another clue to understanding Frege's disapproval of Hilbert. The two protagonists were troubled by two long-standing problems in the philosophy of mathematics: mathematical truth and mathematical existence. Both problems had been accentuated sharply by the discovery of non-Euclidean geometries and the introduction of imaginary and complex numbers. Frege approached these problems via reductionism. The questionable mathematical entities were to be defined in terms of entities of evident existence: one was not simply to postulate their existence (Frege 1884:104–114). On the other hand, the truth of mathematical theorems was to be reduced via an appropriate axiomatization and set of definitions to self-evident or obvious mathematical truths. Hilbert, by contrast, placed no importance on the reduction of one theory to another. Indeed, the closing paragraphs of "Über den Zahlbegriff" imply that reductionism is an unfruitful approach to the problems of truth and existence as they concern the theory of real numbers.

Consistency Proofs

Taking his own axioms to be self-evident, and believing that it is impossible for a genuine axiom to be false, Frege found consistency proofs superfluous. Thus, he wrote to Hilbert:

> It follows from the very truth of the axioms that they do not contradict each other. That requires no further proof. [Frege 1941:409]

Hilbert's reaction was dramatic:

> As long as I have thought, written and lectured about these matters, I have always declared oppositely: if arbitrarily postulated axioms do not contradict each other with their collective consequences, then they are true and the things defined by means of the axioms exist. That, for me, is the criterion of truth and existence.

He also added that he had used this method to establish the existence of the set of all the real numbers and the nonexistence of the set of all Cantorian cardinals (Frege 1941).

In reply, Frege conceded that the question of the consistency of Hilbert's "axioms" could be raised because they are actually defining conditions. However, he added that he knew of no other way of establishing the consistency of a set of conditions except by producing something that satisfies them. As a consequence, Hilbert's criterion would be useless. On the other hand, supposing that there is some other way of establishing the consistency of a set of conditions, Hilbert's method would give rise to another version of the ontological proof: For if it could be shown that the conditions

x is an intelligent being
x is omnipresent
x is almighty

[*115*]

are consistent (without finding an object x satisfying them), then it would follow that an almighty, omnipresent, intelligent being exists (Frege 1941:417). Thus, Frege concluded:

> I cannot accept such a manner of inferring from consistency to truth. You [Hilbert] also probably did not mean it to be this way. In any case, a more exact formulation seems necessary. [Frege 1941:418]

Hilbert's words, taken literally, are open not only to Frege's objection but also to the objection that there are consistent but mutually incompatible axiom systems—both of which, therefore, cannot be literally true of one and the same space. As Hilbert and Frege were well aware of the existence of non-Euclidean geometries, it is hard to believe that Hilbert intended his position to be taken literally. As we saw previously, his other writings at the time suggest that he may have been developing another important idea, to wit: Each axiom set characterizes a class of mathematical structures, but "incompatible" axiom sets treat disjoint structures. The consistency of an axiom set *is* all there is to the mathematical existence of such structures.

Frege's criticism of Hilbert's method is not flawless either. Consider the open sentences

> x is horselike
> x has a single horn.

Let us suppose that they are simple. Frege supposed that the only model theoretic means to demonstrate the joint consistency of these conditions is to show that an object exists which satisfies them, but this is not necessary because it suffices to show that the quantificational schema "$Fx \cdot Gx$" is consistent. And that can be done by assigning new interpretations to "F" and "G" and "x." As this is the very method implicit in Hilbert's consistency proof for geometry, one can understand why he did not reply to Frege's objection that his method could not be applied.

Despite the debatable assumptions behind his objection to consistency proofs, Frege's suspicions concerning the priority of existence and truth to consistency appear correct. As we saw in our earlier discussion of Hilbert, he came to realize that the existence of some mathematical domain and the truth of some statements about it are needed ultimately to establish consistency.

It is unfortunate that Frege and Hilbert did not discuss the consistency problem in greater depth. Possibly, they would have anticipated by several decades some of the most fundamental results of mathematical logic.

Summary and Evaluation

As we have seen, Frege developed the technical foundations for the two forms of the deductivist account of mathematics. According to the *first option,* mathematics is in the business of establishing results in pure logic. These can take the following forms: (1) "$A \supset T$" is logically valid (logically provable); and (2) T is a logical consequence of A (T is logically derivable from A). Here A stands for a quantificational schema diagramming the supposed axioms of a given mathematical theory and T stands for a quantificational schema diagramming the supposed theorem of the theory. Of course, the word "diagramming" is somewhat misleading here, since it presupposes that there is something to be diagrammed, whereas on the deductivist account the "diagrams" are all there is to mathematics. (Let us note in passing that Frege cannot admit approach (2); for on his view quantificational schemata do not have logical consequences—only meaningful statements do.) The second pair of alternatives is the more attractive of the two because by talking in terms of logical consequence or derivability, we can make room even for mathematical theories that have infinitely many axioms and noneffective proof procedures.

The deductivist has a *second basic option* which makes his view appear more in line with traditional mathematics than the first purely logical option. On the second choice, he views a mathematical theory as studying the properties of all structures satisfying certain defining conditions, but he never makes use of the assumption that such structures exist. The deductivist applies the approach of abstract algebra to all mathematical theories. He takes the number theorist to be studying the properties of all structures that satisfy the Peano axioms, the set theorist to be carrying out the same activity with respect to various axiom systems for set theory, and similarly for the geometer, topologist, and analyst.

Unlike the deductivist's first option, this option requires the language of set theory as a basic framework for the mathematician. For it defines each mathematical theory as the study of a certain kind of structure, and the concept of structure is usually defined in set-theoretic terms, unless, of course, one takes Frege's approach and uses second-order logic as the basic framework. (Furthermore, there are grounds for viewing this as set theory in disguise.) However, although the deductivist may use set theory as his basic framework for formulating mathematics, he need not, provided that he is careful in his formulations, make use of any of the existence assumptions of set theory. This would permit him to remain an agnostic with respect to the existence of mathematical structures. Given this, there are definite advantages to choosing the set-theoretic approach rather than the quantificational schema approach. The set-theoretic approach gives mathematics a linguistic framework which is referential—or better, potentially referential—and thus agrees with the prima facie referential character of mathematical language as used by practicing mathematicians. Also, the set-theoretic approach nicely explains the wide use of set theory by mathematicians as a framework system and as a tool for defining new mathematical entities.

Hilbert seems to have been aware of several of the philosophical advantages of the deductivist approach. It can rid mathematics of ontological presuppositions while retaining its apparent descriptive character. He saw that it reduced the epistemology of mathematics to that of logic, and accounts for the centrality of proof in mathematics. It does not force decisions between mutually incompatible mathematical theories, and it permits a great deal of freedom in the construction of new ones. Moreover, it appears to account nicely for the applicability of mathematics, both potential and actual; for when one finds a physical structure satisfying the axioms of a mathematical theory, the application of that theory is immediate. Finally, some parts of mathematics, such as group theory or topology, which are explicitly conceived as abstract characterizations of the structures dealt with by traditional mathematics, make best sense when interpreted along deductivist lines. Thus, it is not difficult to take the extra step of interpreting all mathematical theories along these lines. In view of this battery of considerations in favor of deductivism and Hilbert's profound influence, one can see why it has gained so much popularity among contemporary mathematicians.

The consistency problem was a central issue in the Frege-Hilbert controversy, and rightly so. Consistency remains a nagging issue for the deductivist. An inconsistent mathematical theory is trivial from the mathematical point of view; since any sentence in its language can be derived from its axioms all the difficult labor mathematicians put into proving theorems in a theory can be bypassed completely once an inconsistency has been discovered. Furthermore, these theories have no models. Therefore, the class of structures defined by an inconsistent mathematical theory is simply the empty class. As a result, all inconsistent mathematical theories define the same class of structures. So, if all mathematical theories were inconsistent, there would be no point in developing alternative mathematical theories. Thus, to account for the depth and diversity of mathematics, the deductivist must assume that the great majority of mathematical theories are consistent. This assumption is not unreasonable, especially if it is based upon a belief in mathematical reality and truth which will vouchsafe the consistency of mathematical theories. But what does the deductivist have to fall back on?

In response, a deductivist might argue that since consistency is a mathematical question, it, too, must be treated deductivistically. Thus, the assertion that a given axiom set is consistent must itself be construed as conditional upon a background theory with respect to whose truth the deductivist can remain agnostic. *Within* mathematics we need no more absolute consistency than this, as the discussion of Putnam's paper will reveal. But how can the deductivist leave it at that? For if he is to believe that even some mathematical theories are nontrivial, he must commit himself to their unconditional consistency and thereby to some unconditional mathematical truths.

Hilary Putnam's Deductivism

In addition to the consistency question just discussed, other modern developments have militated against deductivism. One of these is the development of model theory as a branch of mathematical logic. In model theory there are many theorems that do not assert the unconditional existence of mathematical structures. For example, the Löwenheim-Skolem theorem states that if a formal system has a model at all, it has a countable model. But there are also many theorems that do assert the existence of models: for example, the theorems asserting that there is a nonstandard model for first-order number theory. Such assertions seem at first sight to be incompatible with deductivism.

An examination of any modern text in abstract algebra also seems to show that the deductivist apparently is wrong in his assumptions that abstract algebra remains agnostic with respect to the existence of algebraic structures. For, in such texts one is constantly given examples of groups, rings, integral domains, and so on. The texts do not say that if such and such a mathematical structure exists, then it contains as a substructure, say, an integral domain; they state instead that such and such is an integral domain. Furthermore, because these also describe special properties of finite groups and rings, they cannot be restricted to deducing consequences from just the axioms of group theory or ring theory that characterize both finite and infinite groups and rings.

Gödel's famous underivability theorems have also been used to discredit deductivism. For many have taken them to imply that the domain of mathematical truth extends beyond what the deductivist can countenance.

In a study of Bertrand Russell's mathematical philosophy, Hilary Putnam responded to these objections to deductivism and to a further objection that led Russell to reject the deductivist definition of mathematics in his *Principles of Mathematics* (Russell:4–9). I will take up Putnam's replies in the order in which I have presented the objections, and leave the introduction of Russell's objection for the end of my discussion.

The critique of deductivism with which Putnam is concerned is based upon several putative mathematical facts and the claims that deductivism cannot account for them. Putnam uses two strategies. Either he argues that insofar as mathematics makes any use of these facts, they can be interpreted within the deductivist framework; or he claims that the antideductivist interpretation of these facts is fallacious. Consider the problem offered by the existence of model theory. Model theory is a mathematical theory. It proceeds from axioms which, although simply left as understood in most writings in model theory, can be made explicit. Thus, model theory can be formulated as an axiomatic mathematical theory along the same lines as, say, group theory or number theory. When it is so formulated it will be seen to be a branch of axiomatic set theory. Once model theory is thus formulated, the categorical existence statements of model theory can be thought of as hypothetical in the following sense. Any theorem of informal model theory that states that a certain structure is a model for a certain set of axioms can be viewed as asserting that in any structure which satisfies the axioms of the model theory there is a substructure which satisfies the axioms of the theory in question. Indeed, even the statement that there is a standard model for number theory succumbs to this approach. For let M be a structure satisfying the axioms of model theory and let S be the statement

(1) There is a ω-sequence which is a model of number theory.

S states that a standard model for number theory exists. Now relativize S to M; that is, restrict the range of its quantifiers to elements of M. Let this be S_M. Then a deductivist substitute for S is

(2) If M satisfies the axioms of model theory, then S_M holds.

A nonstandard model of number theory relative to M can be defined as a model of number theory relative to M which is not isomorphic (relative to M) to a standard model (relative to M) of number theory. A similar approach can be taken in defining the notion of standard model for set theory. Putnam argues, correctly, I think, that all uses of the terms "standard" and "nonstandard" in the mathematical theory of models can be accounted for on this relativistic basis (Putnam 1967a:283–285).

Putnam takes a similar approach to the assertions in mathematical texts that certain structures are examples of certain kinds of mathematical entities. For example, consider the claim found in many basic algebra texts that the integers modulus thirteen are a ring. This situation is handled easily by Putnam by arguing that an informal model theory is implicit in these texts. With this understanding, the claim that the integers modulus thirteen form a ring can be rephrased as follows: Given any structure M satisfying the axioms of model theory, there exists a substructure formed from the integers of M which can be defined as the integers modulus thirteen of M and which satisfies the axioms of ring theory. In like manner, Putnam can deal with the study of finite rings, groups, and so on, found in algebra texts. A finite group, for example, is taken by Putnam to be a substructure of a structure M satisfying the axioms of model theory, which also satisfies the axioms of group theory and which is finite relative to M. In this way the practice of algebraists can be accounted for without having to postulate the existence of finite or infinite algebraic structures except as relative to the existence of some structure satisfying the axioms of model theory. Nor is this a particularly farfetched account. The algebraist who introduces examples satisfying the axioms of, say, ring theory and then proceeds to make statements about these examples very frequently makes use of facts and methods of reasoning that are not contained even implicitly in the axioms of ring theory. Putnam's approach nicely underwrites these additional inferences and claims. (Putnam also considers, as an alternative response for the deductivist, treating mathematical induction as a primitive mode of deduction—but one whose use is restricted to finite structures and characterizes them. This response seems to me less "clean" than the other one (Putnam 1967a:283–285).)

The reader may have been surprised to notice that Putnam relativizes his discussion of finite mathematical structures to structures satisfying the axioms of model theory. One would certainly have thought that the term "finite" could be given an absolute definition. But, in fact, the discovery of nonstandard models for number theory and set theory has shown that even the distinction between the infinite and the finite must be made relative to a given model for set theory. Since this point can be demonstrated briefly through a fairly elementary argument, it is worth sketching the proof.

Suppose that we are given a formal system for set theory in which we want to introduce an absolute definition of the property of being a finite set. To fix our ideas, let us assume that the system of set theory is Zermelo-Frankel set theory (ZF). We usually proceed to define a class a as finite if and only if there is a natural number n such that a can be put into one-to-one correlation with the set of natural numbers up to n. Here the notions of natural number, one-to-one correlation, and the set of natural numbers up to a given number are defined in the usual way. We now extend the notation of ZF by adding to it a new individual constant, "δ," and extend its axiomatic basis by adding the following axioms: δ is a number, $\delta > 0, \delta > 1, \delta > 2, \ldots$. Suppose that ZF has a model. Then consider any finite subset of the axioms for the augmented system. Interpret them in the original ZF model as follows. Assign the old notation its meaning in the ZF model and interpret "δ" by finding the first numeral not in the finite list of axioms selected and assigning "δ" the number in the original model designated by that numeral. For example, if only "$\delta > 0$," "$\delta > 1$," "$\delta > 2$,"..., "$\delta > 99$" are included in the list, then simply interpret "δ" as 100. All the axioms involving "δ" will then become true, since 100 is greater than 0, 1, through 99. This shows that if the original system has a model, then every finite subset of the axioms of the augmented system has a model. We can conclude, from the compactness theorem, that the augmented system has a model. Yet this new model must be also a model for the original Zermelo-Frankel system. Furthermore, in this model is a natural number greater than $0, 1, 2, \ldots$. Thus, if we consider the set of natural numbers up to δ, we see that it is a set that can be put into one-to-one correlation with a set of natural numbers up to a given natural number, for it can be put into one-to-one correlation with itself. Thus, according to our definition, we have a set in this new model which is finite from the perspective given to us in that model, but infinite relative to the original model. (The compactness theorem states that if every finite subset of a set of first-order sentences has a model, then so does the set.)

At this point, one might object that the inability of the deductivist to make an absolute distinction between the finite and infinite clearly demonstrated the bankruptcy of the deductivistic approach. For does not our experience with natural objects furnish countless examples of finite processes and finite sequences? Can we not select a standard example from which we can define the notion of finite? For example, can we not define a set to be finite if it can be counted in a finite amount of time, allowing a fixed time interval to be given for each step in the counting? (This would avoid the possibility of counting infinitely many elements in a finite amount of time by using smaller and smaller time intervals as we count more and more numbers.) We could then define a finite time interval as one having both a first and a last moment.

Putnam answers this objection with two considerations. The first is that such a course is methodologically incorrect, because it defines a mathematical concept in physical terms and thus destroys the independence of mathematics from physics. Second, the proposal does not guarantee the existence a priori of a standard model for number theory because basic changes in our concept of time could permit an infinite time interval to have both a beginning and an end. Putnam concludes that deductivism "can do justice to talk of 'standard' and 'nonstandard' models *in so far as such talk does any work*" (Putnam 1967a:287).

Putnam turns next to consider antideductivist arguments from Gödel's theorems. He first presents the following argument:

> Suppose I "accept" a set theory T as "correct" in the sense that all the number-theoretic statements provable in T are *true* (in *the* standard model, assuming, for the moment, the "Platonistic" attitude towards standard models which we have just criticized). Then the further statement Con (T) which expresses the consistency of T is one I should *also* accept, since it must be true if T is consistent. Thus we have found a sentence—namely Con (T)—which I have as good grounds for accepting as I have for accepting T itself, but which cannot be proved in T.

Putnam then refutes this argument by claiming that our grounds for accepting "the added axiom Con (T) as preserving consistency" are as good as those for believing that T has a standard model. Thus, it does not follow automatically that when I have good grounds for accepting T, I also have good grounds for accepting Con (T) (Putnam 1967a:287–288).

This argument and Putnam's critique of it are very difficult to follow. The argument itself even appears to be valid! For if all number-theoretic theorems of T are true, then T must be consistent and Con (T) must be consistent with T. For otherwise, one or more false number-theoretic statements would be theorems of T. So what we are supposed to accept concerning the correctness of T already implies both that T + Con (T) is consistent and that Con (T) is also correct (i.e., true in the standard model).

[*124*]

I suspect that Putnam was attempting to present some sort of Gödelian argument that mathematical truth outruns axiomatic proof, which would be problematic, at least initially, for a deductivist. But I see no problems for the deductivist emanating from Gödel's theorems. If someone uses the theorems to challenge the deductivist, he can in turn ask how the antideductivist "sees" that the Gödel sentence is true. If the antideductivist responds with a mathematical argument—the standard one being to the effect that the Gödel sentence is true in the standard model—then the deductivist can extract from the premises of the argument a background axiom set upon which the Gödel sentence is conditional. On the other hand, if the antideductivist cannot present a mathematical argument for the Gödel sentence, the deductivist simply can ignore his claim to see that it is true. In short, Putnam should have used the same reply that he used earlier; insofar as the Gödelian claims do any work in mathematics, they can be treated along the same lines as any other mathematical claim.

I shall now examine some of Putnam's views on Russell's abandonment of deductivism. For convenience, I shall not follow Putnam to the letter on matters of technical exposition.

Frege and Russell independently arrived at an analysis of counting according to which a class has n members if and only if it can be put into one-to-one correlation with the class of numbers less than n. For example, the class of planets has nine members, since the planets can be put into one-to-one correlation with the numbers 0 through 8. A one-to-one correlation is a relation satisfying certain conditions, which can be defined within first-order logic with identity in well-known ways. However, although the property R *is a one-to-one correlation* can be defined within first-order logic, the assertion that there is a one-to-one correlation R between two sets is an assertion of set theory or higher-order logic. This led Russell to conclude that applied mathematics, at least, requires the recognition of mathematical entities such as sets, functions, and relations. He inferred from this that applied mathematics shows the untenability of deductivism (Russell:VI).

Putnam considers a way of avoiding this conclusion. He suggests taking the assertion that there are nine planets as equivalent to the assertion: If M is a model of set theory, then there is a one-to-one correlation in M between the planets and the numbers in M less than the number nine of M. In other words, Putnam considers taking the same hypothetical approach to applied mathematics as deductivism takes to pure mathematics. He immediately rejects this, however, on the grounds that if there is no model for set theory, then all the statements of applied mathematics formulated in the language of set theory will be true by virtue of having a false antecedent. On the proposed analysis, if there is no model for set theory, not only will it be true that there are nine planets, but it will be true also that there are 12 planets! Putnam, like Russell, agrees that the problem of applied mathematics is the most important consideration against deductivism. But he still thinks he can dispose of it (Putnam 1967a:287, 291).

Putnam next notes that, as both Frege and Russell knew, the statement that there are nine planets is not only equivalent to a statement of set theory but is also equivalent to a statement of first-order logic with identity. Indeed, for each finite n the quantifier, "there are $n x$'s" or "$(\exists n x)$" may be defined in first-order logic with identity according to the following recursive scheme.

$$(\exists 0 x) F x \equiv -(\exists x) F x$$
$$(\exists n + 1 x) F x \equiv (\exists y) (F y \cdot (\exists n x) (F x \cdot x \neq y)).$$

Thus, the mere truth of a statement such as "there are nine planets" need not commit us to mathematical structures, since it can be formulated in first-order logic. Consequently, Russell's argument for rejecting deductivism is fallacious (Putnam 1967a:292).

Putnam then attempts to show by considering further examples that applied mathematics will not furnish other cases that will commit us to the existence of mathematical structures. He restricts his attention to simple counting inferences such as the following:

(1) No cat is owned by both John and Bill.
(2) John owns six cats.
(3) Bill owns three cats.
(4) Therefore, nine cats are owned by John or Bill.

Putnam observed that this inference can be carried out within first-order logic with identity as long as we make use of the numerical quantifiers defined according to the scheme given above. Next he claims that according to the Frege-Russell approach, the arithmetical statement ''6 + 3 = 9,'' for example, is equivalent to the set-theoretic statement ''for any classes α and β, if α has six members and β has three members and α and β are disjoint, then $\alpha \cup \beta$ has nine members.'' This set-theoretic statement can be viewed as the law underwriting the counting inference given above, just as the logical schema ''$(p \cdot (p \supset q)) \supset q$'' can be viewed as the law underwriting *modus ponens*. Putnam proposes that we view ''6 + 3 = 9'' as an explanatory principle for the counting inference in first-order logic rather than an essential premise (Putnam 1967a:288–289).

(Putnam's discussion of the Russellian analysis of ''6 + 3 = 9'' needs some correction. The set-theoretic equivalent to ''6 + 3 = 9'' is more like:

(1) γ is the union of some six-membered class α and some three-membered class β disjoint from α if and only if γ has nine members.

Putnam's statement is implied by (1) but implies only the *only if* portion of (1). His approach is too weak to establish arithmetical identities like ''6 + 3 = 7 + 2,'' since it enables us to prove that (6 + 3)-membered classes and (7 + 2)-membered classes are nine-membered but not to show that (7 + 2)-membered classes are (6 + 3)-membered.)

Putnam next raises the question as to how mathematics can be useful, given that it is not essential in counting inferences. More particularly, since ''6 + 3 = 9'' is not essential to the inference in which we counted the cats, how do we explain its usefulness? His answer is that theorems of set theory show that statements of numerical attribution can be formulated equivalently in either first-order logic or set theory. If A is a predicate whose extension is the class a, then it is a theorem of set theory that there are n A's if and only if the class a can be put into one-to-one correlation with the numbers up to n. It is also a theorem of set theory that the cardinal number of a equals n if and only if n can be put into one-to-one correlation with the class of numbers up to n. So if we accept set theory, we can carry out counting inferences through the lengthy route of first-order logic or through the shorter route of set theory. Thus, our example can be formulated in set theory as follows:

> (1)   The class of cats owned by Bill and the class of cats owned by John are disjoint.
> (2)   The number of the class of cats owned by John equals 6.
> (3)   The number of the class of cats owned by Bill equals 3.
> (4) $6 + 3 = 9$.
> (5)   Therefore, the number of the class of cats owned by Bill or John equals 9.

This shows that set theory is useful in counting. But how do we know that it is reliable? Putnam notes that the theorems of set theory cited above show that counting through set theory leads to the same results as counting through first-order logic. But if we got a different result by counting through first-order logic than we got by counting through set theory, then set theory would be inconsistent. Putnam concluded from this that our confidence in the reliability of set theory for applications is grounded in our confidence in its consistency (Putnam 1967a:293).

But does the completeness theorem not show us that set theory is consistent if and only if it has a model? And so are we not back in the position in which Russell found himself, namely, that the application of set theory presupposes that it has a model (i.e., that a set-theoretic structure exists)? Putnam claims not. He claims that the consistency of a theory shows only that it could have models, not that it does have a model. Let me quote him:

> There is a clear difference between believing in the actual existence of something, and believing in its *possible* existence: and I am contending that the employment of [set theory] at least in deriving such statements as we have been discussing from each other, does *not* presuppose "Platonism," i.e. belief in the *actual* existence of sets, predicates, models for [set theory], etc., but only presupposes a belief that a structure satisfying the

axioms of [set theory] is possible. What it is to believe that the existence of something is possible, is a philosophical issue which will not be entered into here; however, this much is clear: to think that *there could be* a structure satisfying the axioms of [set theory], in any sense of "could," is *at least* to be confident that no contradictions will turn up in [set theory]. And that is all we need to employ [set theory] in the manner in which I have been describing. [Putnam 1967a:293–294]

To worries that the completeness theorem requires us to admit a model of a consistent theory, Putnam would presumably respond that the completeness theorem is itself a mathematical result and so must be interpreted along deductivist lines as stating: Given the axioms of model theory and given axioms describing the conditions for a structure to be a formal system, if there are any structures that satisfy both these axioms and if the structure that is a formal system is consistent, then it has a model in the structure satisfying the axioms of model theory. Given this formulation of the completeness theorem, our right to use it is conditional upon discharging its assumptions that an appropriate formal system and a structure satisfying the axioms of model theory exist. Moreover, the last assumption opens the door to set theory. Along these lines, we also can give more content to Putnam's notion that the consistency of a theory implies that it *could have a model* by construing this as meaning: *If the existence conditions appropriate to the deductivist version of the completeness theorem are met, then the theory will have a model.*

Nevertheless, I fail to see how this move has gained Putnam much philosophical headway. He has fallen into the deductivist pitfall of believing that a mathematical theory is consistent. This is to give up the agnosticism concerning mathematical existence and truth, at least at the metamathematical level. For reasons similar to those pushing Hilbert toward the recognition of abstract entities, Putnam, too, is driven to countenance at least a denumerable number of abstract expressions which constitute the elements of a formal system for set theory. But these entities, together with other metamathematical assumptions and Putnam's belief in the consistency of set theory, yield a structure of syntactic objects that satisfies the axioms of set theory. This can be proved as an application of the syntactic version of the completeness theorem due to Hilbert and Bernays. The additional metamathematical assumptions, moreover, are considerably weaker than the assumption that set theory is consistent, since their arithmetized correlates can be proved in elementary number theory. Thus, it seems that Putnam is impelled not only toward the admission of abstract entities, but also toward the admission of a model of set theory, that is, toward the very Platonism that he wished to avoid.

Fred Schmitt has suggested to me that Putnam could avoid this trap by moving on to his position presented in "Mathematics without Foundations" (Putnam 1967b). Deductivism can be construed as asserting that the implication between the axioms of a theory and its theorems is (logically) necessary, thereby reducing mathematics to modal logic. But having made this move, one also could construe the theorems and axioms themselves along modal lines. In particular, the statement that a given formal system is consistent could be construed as asserting that it is not *possible* to *inscribe* a derivation of, say, "$0 = 1$" which meets the rules of the system. This would preserve the nontriviality of claims that a theory is consistent while avoiding commitments to abstract entities. This extension of deductivism needs to be spelled out in detail before it can be evaluated. It appears plausible for syntactic metamathematics, since here we are concerned with symbols or their possible inscriptions; but it may be less plausible when applied to ordinary mathematics, where on the face of it we are less concerned with operations that *we* perform. It is also not clear that this program can be carried out in terms of logical modalities. (See Kessler for further discussion of modalized mathematics.)

Regardless of what conclusions we may draw eventually concerning Putnam's view on consistency, his defense of the deductivist account of applied mathematics is still insufficient. For not only do we make numerical assertions in applied mathematics, such as "there are nine planets," but we also make comparative judgments such as "there are fewer planets in the solar system than there are moons." According to the set-theoretic analysis of cardinality that we have considered, the statement "there are more F's than G's is analyzed as equivalent to

[*130*]

"there is a one-to-one correlation of the G's with a subclass of the F's, but there is no one-to-one correlation of the F's with a subclass of the G's." Furthermore, there is no way to paraphrase statements of this form into equivalent statements in first-order logic with identity, so the considerations that led Russell to abandon deductivism are in force once again. Set-theoretic assumptions are essential to Putnam's account of applied mathematics and not simply a convenient auxilliary.

It should be pointed out that my last objection to Putnam's view of deductivism depends upon his analysis of cardinality. My argument need not effect someone who had an alternative analysis. Furthermore, Putnam himself later abandoned deductivism and has decided that difficulties in accounting for applied mathematics are decisive (Putnam 1975:31).

Some General Considerations against Deductivism

The deductivist can, if he wants, present a particularly simple epistemology for mathematics. He need only claim that we know a mathematical statement "if AX $(T) \supset S$" just in case we know that S can be derived from the axioms of T by means of (first-order) logic. Of course, it is open to the deductivist to propose a more extended epistemology. It has been suggested to me, for example, that a deductivist might permit the use of analogies concerning deductions that have already been completed. In fact, neither Hilbert nor Russell nor Putnam considered such extensions, and they would detract from the basic deductivist position. In any case, the consideration against deductivism I shall now offer need not affect more complicated versions, although in fact it has convinced a number of contemporary thinkers—Putnam is one—to turn away from deductivism. The consideration is this: throughout the history of mathematics mathematicians have been convinced completely of results by means of arguments that fall short of deductive proof. Furthermore, if we reject the claim that these mathematicians knew the results, then we would also have to reject very firm knowledge claims in other areas of science. Both Putnam and Mark Steiner give a number of examples of this kind (Putnam 1975:64–69; Steiner:102–108, 136). One that both discuss is an instance in which

Euler arrived at a result by a very sophisticated combination of analogical reasoning, extensive verification of instances of the result through computation, and checks of his reasoning by applying it to results that had already been proved. According to Steiner and Putnam, Euler was absolutely convinced of his result; and any rational being would have been convinced of the result given the same data that Euler had. Thus, Euler was entitled to credit for the knowledge of his result; however, not only did he lack a rigorous proof, he did not even possess the concepts and methods for producing a proof. Steiner also cites the fabulous powers of Ramanujan, an Indian mathematician who, although completely untutored in the standards of Western rigor, produced extremely complicated formulas in the theory of elliptic functions which were proved later by other mathematicians. In addition, we could cite Newton and Leibniz, who arrived at correct results with inadequate methods, which are not capable of direct formalization. Certainly, there are many similar cases in the history of mathematics. These cases are particularly embarrassing to the deductivist, because he claims that a mathematical result always implicitly refers to axioms from which it is supposed to follow. In the cases in question, however, either the mathematicians were unable to cite any axioms or else the axioms they did cite were insufficient to generate the result. As long as we are willing to credit these historical mathematicians with their results, we should be also willing to recognize mathematical evidence other than deductive proof.

A related objection, one Brouwer raised against the formalists, is that the deductivist position on the choice of axioms conflicts with actual mathematical practice. According to the deductivist, it would be perfectly legitimate for mathematicians to make up axiom sets through some random method and then proceed to investigate their logical properties. But mathematics does not proceed in this way. Not only have the new axiom systems historically developed from issues and problems encountered in the study of previous mathematical theories, but also certain consistent axiom systems are not considered worthy of serious mathematical investigation. For example, we would not develop set theory with the negation of the pair-set axiom, although it is possible to set up a set theory in which the negation of the pair-set axiom is consistent with the other axioms. Some logicians have tried the permutations and combinations approach to axiom systems (especially for the propositional calculus), but this has been derided by mathematicians as not being in the proper mathematical spirit. Furthermore, consistent axiom systems, even highly developed ones, have been dismissed by set theorists as being along the wrong lines, as not very fruitful, as pursuing an unclear idea, as not admitting the proper kind of model, and so on. This shows that mathematicians are prepared to find some consistent mathematical systems to be unacceptable. But once one can distinguish between consistent axiomatic systems, one can make room for unconditional truths in mathematics. These are not truths of applied mathematics. They are affirmed by the mathematicians themselves, as when, for example, they say: "Sure, set theory with the negation of the pairing axiom is consistent, but it is not true."

An energetic deductivist might respond to this objection by arguing that although simplicity, elegance, and fruitfulness are reasonable criteria by which to evaluate axiom systems, they bear only upon the aesthetics of mathematics rather than its truth. Thus, while my introduction of mathematical practice shows that deductivism needs supplementation, it fails to show that mathematicians consider truth to be a relevant factor in choosing axioms to study. My response to this would be to argue that since these very factors are usually deemed relevant to determining truth in physics, the deductivist separation of mathematics and physics seems particularly arbitrary.

This brings us to the view the deductivist is forced to take of the language of the working mathematician. According to the deductivist, the sincere affirmations of the mathematician that a certain mathematical structure exists and that certain statements are true are to be viewed as elliptical for longer conditional statements which the mathematician may never have uttered. Furthermore, if we confront a mathematician with the claim that his statements are elliptical and he steadfastly denies this, we are to assume that he simply does not understand the situation as deeply as we do. By contrast, the deductivist does not make the same claims with respect to the language of the physicist, for the physicist is allowed his sincere affirmations that certain physical entities exist. Thus, the deductivist must defend an epistemological and linguistic distinction between mathematics and physics to justify this separate account of talk of mathematicians and physicists. But how can he do this without running into even deeper philosophical difficulties? For example, he would need to explain why realism is acceptable in nuclear physics but not in mathematics, and he would need to answer Quine's critique of the possibility of drawing an epistemologically significant boundary between these two sciences (Quine 1951; 1962).

The philosophical advantage surely goes to a view that offers a uniform account of the language of science and its epistemology—and this includes the science of mathematics. Not only do such accounts promise to be simpler, but they are also in a better position to deal with the intimate bond between mathematics and the natural sciences. We should not forget that the fundamental laws of physics are expressed in mathematical formulas and that the activities of the theoretical physicists are barely distinguishable from those of the working mathematician. Nor will it do for the deductivist to respond that *all of science* may be interpreted as the study of the logical consequences of axioms and hypothesis. For that would push hypothesis testing, observation, prediction, and inductive reasoning entirely from the domain of (deductivist) science.

As we are discussing the deductivist contrast between mathematics and natural science, it is appropriate to consider the deductivist account of the application of mathematics. Here one would have thought that the deductivist is on the firmest possible ground. This is best seen if we return to Hilbert's form of deductivism, according to which the mathematician is deducing logical consequences from theory forms or sets of quantificational schemata. Unlike the set-theoretic form, this will not require the applied mathematician to recognize the existence of set-theoretic entities before he can apply a mathematical theory. On the theory-form version of deductivism, once one finds a set of truths that have the same logical forms as the axioms of a known mathematical theory, one can immediately apply that theory and use all the logical consequences that have been derived from the axioms by working mathematicians. I have no qualms with this part of the account. Nor do I have serious objections to the account at all, insofar as it concerns simple theories of finite structures. My concern is with the application of sophisticated mathematical theories, such as the theory of real numbers or analysis.

To determine that the axioms for such a theory are satisfied, one would have to be convinced that a fairly complex infinite structure exists. We have ample reason to believe that no such structure is found among purely physical objects. So the existence of such structures will involve the existence of infinitely many abstract entities. On the deductivist account, the question as to whether there are structures satisfying the axioms of analysis is of no concern to the mathematician. But it seems also not to be the concern of the working physicist, insofar as his attention is devoted to physical objects. But if it is not the concern of the mathematician, whom can it possibly concern? I agree with Quine that the deductivist has ruled out the concerns of the traditional mathematician via a verbal tour de force (Quine 1936:76; see also Frege 1903a: sec. 92).

Perhaps the deductivist will say at this point that the physicist is the one who determines whether a given mathematical axiom system should be used within physics, but he need not verify that the axioms are true of a certain physical structure to be justified in making the application. He simply has to be assured that the physical structure sufficiently approximates the mathematical structure or that the axioms are approximately true of the physical structure. But not only do we lack a theory of approximation through which we can compare two theories, but such a theory probably would have to be supported by a very sophisticated mathematical theory which itself would include at least the theory of real numbers. So this attempt to explain the applicability of mathematics seems destined to run into difficulties. Thus, someone, possibly the physicist, mathematical biologist, or economist, seems fated to recognize structures satisfying some mathematical axioms. So what has the deductivist gained? Well, the mathematician at least need not worry about the truth of his axioms, and there will be many axiom systems—those which remain unapplied—whose truth or falsity will be of concern to no one. But this gain pales when we recognize that many difficult questions traditionally treated by the philosophy of mathematics have merely been shifted over to the philosophy of science without a start toward answering them.

Furthermore, in eschewing traditional mathematical truth, deductivism may undercut the epistemology for logic itself—especially if we heed the considerations raised by Brouwer and adopted by Hilbert concerning the reliability of logical inference in mathematical reasoning. The point is this: part of the usual justification for accepting a rule of deductive inference is that following it invariably leads from premises that are true (or accepted or have whatever characteristic goes proxy for truth in one's semantics) to conclusions which also are true (or accepted, etc). But since deductivism provides for no concept of mathematical truth beyond logical truth, it can be sure only of the correctness of the logical rules in their nonmathematical applications. Because such applications presumably are restricted to finite physical structures, the Brouwer-Hilbert doubts about the use of logic in application to abstract infinite structures—even those only hypothetically existing—remain unanswered. The deductivist must, therefore, fall back upon a stronger conception of logical truth, such as conventionalism or Peirce's view that the mathematician is able to *intuit* the consequences of his axioms (Peirce). The conventionalist account must deal with Quine's objection that logic itself is needed to derive the consequences of any (recursive) set of conventions used to set up logic (Quine 1936). On the other hand, Peirce's approach again introduces a mysterious faculty into mathematical epistemology, thereby undermining deductivism's major gain.

In conclusion, deductivism is a powerful and appealing philosophy of mathematics, for it allows mathematics a referential language while avoiding commitments to abstract entities, and it also simplifies its epistemology. However, deductivism leaves a number of philosophical loose ends. There is the concern about its ability to account for mathematical practice and the problem of justifying the linguistic distinction that it draws between mathematics and physics. Furthermore, there is the Frege-Quine complaint that the mathematical axioms that must be accepted in applying mathematics have been shunted into a no-man's-land between mathematics and the other sciences. Finally, there is the difficulty that deductivism may be unable to present a satisfactory epistemology for deductive reasoning itself. These objections need not furnish a definite refutation of deductivism, but since they remain unanswered they represent a difficult and pressing challenge for present and future deductivists.

[*136*]

Mill's Empiricism

Frege was fond of deriding John Stuart Mill's "pebble and ginger-bread" account of arithmetic. He devoted as much space in the *Grundlagen* to its refutation as he does to any other view, and his later polemical works are spiced with digs at Mill. (See, for example, Frege 1894:315.) Frege's criticism focused upon Mill's account of the natural numbers and the axioms and postulates of arithmetic. It quite correctly exposed a number of basic faults and confusions in Mill's doctrines, thereby effectively showing it to be an unworkable approach to mathematical foundations. Nonetheless, Frege's critique is unsympathetic; and Frege wrongly concluded from it that all empiricist approaches are unworkable. Thus, my aim in this chapter is to give a fuller exposition of Mill's philosophy of mathematics as well as to evaluate Frege's critique and to determine whether a more workable account can be mined from Mill's views. I will open with an exposition of Mill's views.

Mill's View

Mill's doctrines can be summarized as follows: mathematics is known inductively from experience. Mathematical reasoning—indeed, all so-called deductive reasoning—is in reality inductive. The general axioms of mathematics are inductions derived from experience, and its existential postulates assert matters of physical fact. Moreover, there are no special mathematical objects, for mathematics is simply a very general theory of ordinary physical objects.

Mill's theory of definitions is a good place to begin the examination of the details of his view, because he was especially concerned to refute the view that mathematics is an a priori science which can ultimately be reduced to a series of definitions and their logical consequences. He accomplished this by arguing that most—and certainly the important—definitions of mathematics are more than simple meaning conventions. Mill recognizes two types of definition. The first consists of explanations or stipulations of meanings. Mill's example is the following definition: A centaur is an animal the upper half of whose body is the body of a man and the lower half of whose body is the body of a horse. This definition, he claimed, simply fixes the meaning of the word "centaur." On the other hand, definitions of the second type, which includes typical mathematical definitions, are not merely meaning stipulations but also covert assertions of fact (Mill:Bk. I, chap. VII, sec. 4). Mill gave this example: A triangle is a rectilineal figure with three sides. According to Mill, this definition contains two assertions: that rectilineal figures with three sides exist and that such figures are to be called triangles. So we have in the definition not only a stipulation concerning the word "triangle," but also an implicit existence assertion. The latter, he claimed, is why the definition is so useful in geometrical reasoning and why without its existence assertion, no significant geometrical theorems about triangles could be deduced from it.

Mill also held that definitions in arithmetic contain implicit existence postulates. This is a point Frege was most critical of, so I shall quote Mill:

> We may, if we please, call the proposition, "three is two and one," a definition of the number three, and assert that arithmetic, as it has been asserted that geometry, is a science founded on definitions. But they are definitions in the geometrical sense, not the logical; asserting not the meaning of a term only, but along with it an observed matter of fact. . . . we call "Three is two and one" a definition of three; but the calculations which depend upon that proposition do not follow from the definition itself, but from an arithmetical theorem presupposed in it, namely, that collections exist, which while they impress the

senses thus ⃝⃝, may be separated into two parts, thus ∞ ⃝. This proposition being granted, we term all such parcels Threes, after which the above-mentioned physical fact will serve also for a definition of the word Three. [Mill:Bk. II, chap. VI, sec. 2]

Since the arithmetical definitional postulates assert matters of fact, they are justified by making the appropriate observations and verifying that such collections exist (Mill:Bk. II, chap. VI, sec. 2; Bk. III, chap. XXIV, sec. 5). However, Mill did not take geometrical existence postulates to assert matters of observable fact, because he did not believe that there are any true points, lines, planes, triangles, circles, and so on, to be met in experience. For any point that we experience will have some magnitude; any line that we draw or see will have some breadth and cannot be perfectly straight; any triangle will not be a perfect triangle; and so forth. Nor would Mill permit geometrical objects to exist as *possibilia* or conceptual idealizations, since none of our mental images contains perfect lines, circles, and so on (Mill:Bk. I, chap. VIII, sec. 6; Bk. II, chap. V, secs. 1–2). Instead, his position was that the existential postulates and other axioms involving geometrical objects are *approximately* true of actual physical objects. Just as in physics it is not practical for us to consider every force, no matter how small, which acts upon a missile when determining its trajectory, so in reasoning about ordinary spatial objects we neglect their irregularities—for simplicity's sake—and derive conclusions that are approximately true of them. In specific applications of geometry it may be necessary to "correct our conclusions by combining them with a fresh set of propositions relating to the aberration" of the real objects from the geometrical ideals (Mill:Bk. I, chap. V, sec. 1).

This view of geometrical objects quite naturally leads Mill to endorse deductivism with respect to geometry: "When, therefore, it is affirmed that the conclusions of geometry are necessary truths, the necessity consists in reality only in this, that they correctly follow from the suppositions from which they are deduced" (ibid.). Thus it would seem that Mill should distinguish pure geometry, in which we develop the logical consequences of postulates and axioms suggested to us from experience, and applied geometry, in which we determine the extent to which these hypotheses and their consequences are true of specific physical objects.

Unfortunately, Mill's view is not that simple or that attractive. First, his doctrine of approximate truth is rendered confusing and less plausible by his claim that the geometrical properties we postulate do not

> involve anything which is distinctly false, and repugnant to its real nature: we must not ascribe to the thing any property which it has not; our liberty extends only to slightly exaggerating some of those it has. . . . that the hypotheses should be of this character is, however, no further necessary, then inasmuch as no others could enable us to deduce conclusions which, with due corrections, would be true of real objects. [Mill:Bk. ii, chap. v, sec. 2]

There are two problems here: (1) if a line is not exactly straight, then to call it straight is to ascribe to it a property it does not have—and exaggeration is a type of falsification; and (2) there is no unique geometrical description of, say, a roughly triangular object which will enable us to reason about it—it can be idealized as either a Euclidean or a non-Euclidean triangle. We can help Mill out of both these difficulties by invoking a pragmatic concept of simplicity. Suppose that, in order to purchase seed, we need to estimate the area in a field that is more trapezoidal than rectangular. It is usually simpler to calculate the area of a rectangle enclosing the field and purchase extra seed, since the costs both in terms of money and plant competition are negligible. On the other hand, if land of the same dimensions were being sold by the square foot, as it is in Honolulu, then a more exact description than even the trapezoidal one would be used. It would be *possible* to use the rectangular one, but the attendant corrections necessary to get an adequate estimate of the area involved would be enormously complex. In this case it is simpler for a surveyor to determine the polygon enclosing the land, and he needs this for the deed description in any case. Similarly, in most applications it is easier to treat a triangular object as Euclidean—but not always, as contemporary astronomy teaches us.

The second complication in Mill's view is that he recognized no genuine deductions. All inferences, for him, reduce to simple induction from particular facts to particular facts. Consider, for example, the purportedly deductive inference:

> All humans will die
> I am a human
> Therefore, I will die.

According to Mill the true inference is from the second premise, the particular fact that I am human, to the particular conclusion that I will die. The deductive form of the inference given above and the use of the universal premise is simply a convenient device for summarizing our inductive inferences from particulars to particulars.

> From instances which we have observed, we feel warranted in concluding, that what we found true in those instances, holds in all similar ones. . . . We then, by that valuable contrivance of language which enables us to speak of many as if they were one, record all that we have observed, together with all that we infer from our observations, in one concise expression. . . .
>
> When, therefore, we conclude from the death of John and Thomas, and every other person we have ever heard of in whose case the experiment had been fairly tried, that the Duke of Wellington is mortal like the rest, we may, indeed, pass through the generalization, All men are mortal, as an intermediate stage; but it is not in the latter half of the process, the descent from all men to the Duke of Wellington, that the *inference* resides. The inference is finished when we have asserted that all men are mortal. What remains to be performed afterwards is merely deciphering our own notes. [Mill:Bk. II, chap. III, sec. 3]

Lest we protest that then the inference is, indeed, from particulars to a generalization, and thence to new particulars, Mill continues in the next section with:

> General propositions are merely registers of such inferences already made and short formulae for making more: The major

premise of a syllogism, consequently, is a formula of this description; and the conclusion is not an inference drawn *from* the formula, but an inference drawn *according* to the formula; the real logical antecedent or premise being the particular facts from which the general proposition was collected by induction. [Mill:Bk. II, chap. III, sec. 4]

Thus, for Mill the true form of my inference would be

$a_1, a_2, \ldots, a_n$ are humans who have died, I am a human, thus I will die.

where $a_1, a_2, \ldots, a_n$ are those human beings whose deaths are known to me and upon which I base my inference that I will die.

Mill also argues that all syllogistic reasoning, and thus all deductive reasoning, can be reduced to this sort of induction (Mill:Bk. II, chap. III, secs. 1–2). We need not explore all the linguistic, logical, and epistemological difficulties which this account of deduction entails. Mill did attempt to cope with them, but in the light of our current knowledge of the deductive logic required for mathematics, they are bound to be unsurmountable. For completeness, however, we should examine how Mill hoped to fit this to his "deductivistic" account of geometry.

Mill argues that in any geometrical argument we are reasoning about a particular (idealized) figure and so do not require general axioms or postulates but only certain of their instances. Nevertheless, the reasoning involved is paradigmatic and permits us to induce the general hypothetical theorem. Thus, we may have both our inductivist account of geometrical reasoning and our deductivist account of the necessity of its theorems. To quote Mill:

> What assumption, in fact, do we set out from to demonstrate by a diagram any properties of the circle? Not that in all circles the radii are equal, but only that they are so in circle *ABC*. As our warrant for assuming this, we appeal, it is true, to the definition of a circle in general; but it is only necessary that the assumption be granted in the case of the particular circle supposed. From

this, which is not a general but a singular proposition, combined with other propositions of a similar kind, some of which *when generalized* are called definitions, and others axioms, we prove that a certain conclusion is true, not of all circles but of a particular circle *ABC*; or at least would be so, if the facts precisely accorded with our assumptions. The enunciation, as it is called, that is, the general theorem which stands at the head of the demonstration, is not the proposition actually demonstrated. One instance only is demonstrated: but the process by which this is done is a process which, when we consider its nature, we perceive might be exactly copied in an indefinite number of other instances; in every instance which conforms to certain conditions. The contrivance of general language furnishing us with terms which connote these conditions, we are able to assert this indefinite multitude of truths in a single expression, and this is the general theorem. [Mill:Bk. II, chap. III, sec. 3]

The attractiveness of these thoughts wanes, however, when we remember that each step in a proof about an individual circle is an induction from a particular to a particular. This also detracts from Mill's theory of approximate truth, since that makes much better sense when we think of ourselves operating deductively with idealized hypotheses whose consequences are checked for their fit with experience.

It might be objected that I have been too quick in dismissing the deductivist themes in Mill's work. I certainly do not wish to deny that he frequently asserts that whatever necessity is to be found in mathematics is the relative necessity the theorems have with respect to the axioms and definitional postulates. Yet most philosophies of mathematics grant at least this much necessity to mathematics. Part of what distinguishes Mill from the true deductivist—as I am using the term—is that the latter holds that the deductions in question yield a priori knowledge and do not depend upon the applicability of their assumptions for their scientific status. Furthermore, as we shall see shortly, Mill's other theses separate him even further from deductivists.

Before leaving Mill's view on definitions, I should state that Mill was quite correct in claiming that mathematics cannot be reduced to mere identities and abbreviatory definitions, and that both geometry and arithmetic depend upon existential postulates. But, at least as far as modern presentations of arithmetic and geometry are concerned, his examples and attempts to prove that definitions do contain existential postulates through an implicit use of the doctrine of existential import (Mill:Bk. I, chap. VIII, sec. 5) are not quite on the mark. Thus, in the definition "3 = 2 + 1" no existential postulate is asserted. But in treating "2 + 1" as a full-fledged singular term we do presuppose that the numbers are closed under addition of one, which involves an existential theorem, at least in a formulation of arithmetic whose primitive notation contains only predicates.

Let us now turn to a discussion of Mill's doctrines concerning the proper (i.e., nondefinitional) axioms of mathematics. As it turns out, Mill could not be a deductivist, even if we ignore his reduction of deduction to induction, because of his view on proper axioms. For these are general truths which are used in the mathematical sciences as bases for mathematical proofs, and are direct, inductive generalizations from observation. It will be useful to follow Mill's exposition and separately consider the geometrical and arithmetical sciences. In each case Mill proceeded by criticizing a prominent theory contrary to his own and then proposed his own account of how axioms are justified through inductive generalization from observations.

In his discussion of geometry, Mill concentrated on the axiom that two straight lines cannot enclose a space. At the time that Mill wrote, the dominant view of this axiom was that it is an a priori truth known by reflection upon the manner in which the mind constructs space. Mill thought that this axiom is amply evident from experience itself, and that there is no need to concoct an a priori justification for it. Mill considered two arguments to the contrary.

The first of these is that we are not convinced of the truth of the axiom by experimentation alone, but arrive at its truth through thought experiments, since in fact empirical evidence for geometrical axioms is not even obtainable. In the case under discussion, for example, we cannot actually follow two intersecting straight lines through their infinite paths to determine by actual observation whether they ever meet again. Mill replied that this is quite true. We cannot carry out the appropriate experiments, and we do indeed verify the axiom by conceiving of straight lines drawn by imagination and imagining ourselves as following them to infinity. *But,* he continued, our justification for taking imagination to be a reliable method for determining the properties of space is dependent upon the inductive generalization that imagination adequately represents spatial reality. Mill claimed that this is what we do when we determine properties of a thing by examining a photograph of it. We have ample inductive evidence that photographs are sufficiently adequate copies of what they depict to enable us to use them to determine certain properties of these things. But just as our use of photographs does not establish that the knowledge obtained from them is a priori, so also the use of imagination in geometry fails to establish that it is an a priori science (Mill:Bk. II, chap. V, secs. 3, 4).

The second argument Mill examined concludes that the axioms of geometry are a priori because they are conceived by us as necessarily true and universal in their application. (This is the Kantian criterion for a priori judgments.) Yet mere observation will confer neither necessity nor universality upon a proposition. Although we may observe many empirical figures that approximate straight lines, all we can conclude from this is that no two observed lines enclose a space. These observations do not furnish us with the conviction that all straight lines have this property or that all straight lines necessarily have this property. Mill rejected this argument by observing that to say that a proposition is necessary is simply to say that its negation is inconceivable, and this is temporally and culturally relative. One may find something inconceivable today and, with further experiments and conceptual development, find it perfectly conceivable and perhaps even true at a later time. Mill cited examples from the history of science which attest to this. He pointed out that when Newton's theory of gravity was first advanced, the Cartesians objected to it as inconceivable because it requires that each physical object is subject to gravitational attraction from every other physical object, even those very remote from it (Mill:Bk. II, chap. V, sec. 5). Similarly, the falsity of Euclidean geometry was inconceivable until non-Euclidean geometry was sufficiently well developed. Of course, before then, one could suggest that the axioms of Euclidean geometry are false, but until a coherent conceptual alternative to Euclidean geometry had been developed, simply saying that one or more of the axioms of Euclidean geometry is false was mere mouthing of words. At that time one could not form a coherent picture of what space would be like if it were not Euclidean. But once an alternative conception had been developed, it became perfectly rational to entertain the hypothesis that one or more of the Euclidean axioms are false. (This point has also been made by Hilary Putnam.)

Frege would have been opposed vehemently to Mill's empiricist account of geometry, since for Frege it is a synthetic a priori science based upon our spatial intuitions. However, Frege did not evaluate Mill's views on geometry, so I shall insert my own comments at this point. Although Mill's transition from the modal notions of possibility and necessity to the psychological notion of conceivability is arguable, I endorse its use in the case of geometry. History has supported his claim amply. However, I find his discussion of the use of imagination and thought experimentation to justify or discover geometric axioms to be confusing and of dubious foundation.

According to Mill, true geometrical points, lines, and planes are not found in the ordinary world of experience or in imagination, but are postulated by geometricians as idealizations of figures that actually are experienced. Thus, I am puzzled by Mill's use of imagination to determine the properties of these figures; for the natural extension of his view would be that their properties are also postulated via extrapolation and idealization. Perhaps Mill made the mistake of conflating imaginary (i.e., hypothetical) figures with mental images. In any case, imagination cannot adequately found geometry. For the geometrical figures that we imagine are themselves not sufficiently pure to guarantee that their properties are Euclidean. (Two non-Euclidean lines *can* enclose an area.) Thus, the inspection of mental images is just as inconclusive with respect to the axioms of geometry as is the inspection of experientially encountered figures.

Furthermore, there is a strong disanalogy between photographs and mental images. Photographs can be compared objectively with each other and with the objects they depict. A combination of a highly developed optical theory and abundant practical experience can be used to determine how enlargement, exposure, lens and film type, focus, and so on, affect photographic images. As a result, we are in a position to fix the limits of the use of photographs to ascertain properties of the objects they depict. We can hardly make such a claim for mental images. Thus, I cannot accept Mill's grounding of geometry in imagination.

Let us now take up Mill's account of the axioms of arithmetic. He begins by considering a view reminiscent of Leibniz's theory, namely, that arithmetic may be reduced via definitions to mere identities. He quickly disposes of this by pointing out that it is "so contrary to common sense that a person must have made some advance in philosophy to believe it: Men fly to so paradoxical a belief to avoid, as they think, some even greater difficulty which the vulgar do not see" (Mill:Bk. II, chap. VI, sec. 2). This cavalier dismissal is mitigated by Mill's extensive discussion of mathematical definitions given in previous sections. Given what he said there, the crucial definitions of the Leibnizian view, such as "3 = 2 + 1," cannot be mere verbal or abbreviatory definitions. Thus, arithmetic cannot be reduced to mere identities. I certainly agree with Mill's insight, and so would Frege—although neither of us would endorse the details of his reasoning.

Mill continued by asserting without real argumentation that the axioms of mathematics are derived by induction from observed experience. To make this claim plausible, he added the following explanation:

> There is in every step of arithmetical or algebraic calculation a real induction, a real inference of facts from facts; and what disguises the induction is simply its comprehensive nature and the consequent extreme generality of the language. All numbers must be numbers of something; there are no such things as numbers in the abstract. *Ten* must mean ten bodies, or ten sounds, or ten beatings of the pulse. But though numbers must be numbers of something, they may be numbers of anything. Propositions, therefore, concerning numbers have the remarkable peculiarity that they are propositions concerning all things whatever, all objects; all existences of every kind known to our experience. All things possess quantity, consist of parts which can be numbered, and in that character possess all the properties

which are called properties of numbers. That half of four is two must be true whatever the word four represents whether four hours, four miles, or four pounds weight. We need only conceive a thing divided into four equal parts (and all things may be conceived as so divided) to be able to predicate of it every property of the number four, that is, every arithmetical proposition in which the number four stands on one side of the equation. Algebra extends the generalization still further; every number represents that particular number of things without distinction, but every algebraic symbol does more; it represents all numbers without distinction. [Mill:Bk. ii, chap. vi, sec. 2]

There are many confusing elements in this passage. To begin with, Mill claims that the word ''ten'' means 10 bodies, 10 sounds, 10 strokes of the clock, and one wants to ask: Which of these does it mean? How can it mean all of them at once? The resolution of this problem is found in Mill's theory of meaning, since he treated number words as predicates and claimed that predicates have a two-part meaning. The predicate ''ten'' *denotes* the particular entities to which it applies (i.e., all parcels or aggregates of 10 physical objects). The word ''ten'' also *connotes* (as the other part of its meaning) the property that all the entities denoted by ''ten'' share and by virtue of which we apply the word to them (Mill:Bk. iii, chap. xxiv, sec. 5; Bk. i, chap. ii, sec. 5).

(It is important to distinguish Mill's theory of meaning from Frege's theory of sense and reference. Frege recognized a two-part distinction in the meaning of a word, but he differs from Mill on several points. On Mill's account a predicate may denote several things, while on Frege's account a predicate *refers* to a single entity called a *concept.* The things to which a predicate applies are, according to Frege, the things that fall under the concept to which the predicate refers and belong to the extension of that concept. The extension of a concept is an abstract entity similar to a class or set. Finally, with each predicate Frege associates an intentional entity called its *sense,* which is similar to Mill's connotation in that one correctly understands the sense of a predicate just in case he knows how to apply the predicate correctly. Frege thus associates three entities with a predicate: its reference (the concept it refers to), its sense, and its extension (the class of things to which it applies). Furthermore Frege, unlike Mill, ascribed both sense and reference to proper names.)

I find Mill's remark about the properties of the number four being transferable to each of its denotations far more suspect than his theory about the meaning of the word "ten." It is difficult to make sense of Mill's claim that four strokes of a clock share all the properties of the number four, especially when we consider its arithmetical properties. For example, four has the properties of being divisible by two and being equal to double its positive square root. Now what sense can we make of saying that four strokes of a clock is double of its positive square root? In Frege's terms, Mill's mistake is this: Mill takes numbers to be properties of aggregates of objects, but then the properties of numbers would be properties of properties of aggregates of objects, and there is no reason to believe that these higher-level properties can be transferred to the lower-level aggregates.

We can make a little more sense of Mill's view by reformulating it as follows: Numbers are properties of aggregates of physical objects, where by an aggregate we mean a sum individual in the sense of Nelson Goodman's calculus of individuals (Goodman). These aggregates are not classes, sets, or other abstract entities. They are concrete individuals with detached parts, like a pack of cards or a case of beer. Numbers are directly observable properties of aggregates, just as spatial arrangement is a directly observable property of certain aggregates or groups. Furthermore, arithmetical operations correspond to physical operations with collections. For example, addition corresponds to grouping two collections together. To quote Mill:

> Every arithmetical proposition; every statement of the result of an arithmetical operation; is a statement of one of the modes of formation of a given number. It affirms that a certain aggregate might have been formed by putting together certain other aggregates, by withdrawing certain portions of some aggregates; and that, by consequence, we might reproduce these aggregates from it, by reversing the process. [Mill:Bk. III, chap. XXIV, sec. 5]

[*149*]

On this view, we could then construe the equation "3 + 2 = 5" as shorthand for something like "3*F*'s + 2*F*'s yields 5*F*'s" and 5*F*'s can be decomposed into 3*F*'s and 2*F*'s. We could then claim that this equation has been established inductively by observing collections of three things and two things and putting them together and decomposing them. Furthermore, if we can characterize numbers generally, then we could have a basis for inductive generalizations that deal with all numbers. Mill thinks that such a property of all numbers can be found:

> What, then, is that which is connoted by the name of number? Of course, some property belonging to the amalgamation of things which we call by the name; and that property is, the characteristic manner in which the amalgamation is made up of, and may be separated into, parts. [Mill:Bk. III, chap. XXIV, sec.5]

Now, given that this singles out a certain class of properties (which is very doubtful), we could then establish general laws of algebra by induction on collections of physical objects and properties belonging to this class. This would involve constructing a law such as "*n* + *m* = *m* + *n*" as "for any collection *C* and any numerical properties *n* and *m*, *nC* + *mC* yields *mC* + *nC* and conversely." Although this clarifies Mill's view somewhat, it is obviously a very crude view and is open to several objections. This will become apparent when we consider Frege's criticisms of Mill.

Another problem with which Mill must deal is that in arithmetic we do not establish nongeneral "laws," such as "3 + 2 = 5," by induction, but rather derive them from a very small set of axioms. Mill's account (as presented up to now) would lead us to believe that there is a unique axiom for each numerical equation. Mill asserts that all properties of numbers are determined by the manner in which they are formed from earlier numbers. Now, the greater a number, the more ways in which it can be formed from earlier numbers. Thus, 2 can be formed only by adding 1 to 1, but 6 can be formed by adding 3 and 3 or 2 and 4 or 5 and 1, and so on. Nonetheless, we can find a single uniform way of generating each number, namely, by generating it from its predecessor by adding 1. Then, given the position of a number in a number series, we can determine how it has been composed from all previous numbers and in this way determine its properties. Now, I quote Mill:

> What renders arithmetic a deductive science, is a fortunate applicability to it of a law so comprehensive as "the sums of equals are equals:'', or (to express the same principle in less familiar but more characteristic language), whatever is made up of parts is made up of parts of those parts. This truth, obvious to the senses in all cases which can be fairly referred to their decision, and so general as to be coextensive with nature itself, being true of all sorts of phenomena (for all admit of being numbered), must be considered an inductive truth, or law of nature, of the highest order. And every arithmetical operation is an application of this law, or of other laws capable of being deduced from it. This is our warrant for all calculations. We believe that 5 and 2 are equal to 7, on the evidence of this inductive law, combined with the definitions of those numbers. We arrive at the conclusion (as all know who remember how they first learned it) by adding a single unit at a time: $5 + 1 = 6$, therefore $5 + 1 + 1 = 6 + 1 = 7$: and again 2 equals $1 + 1$, therefore, $5 + 2 = 5 + 1 + 1 = 7$. [Mill:Bk. III, chap. XXIV, sec. 5]

Frege was to observe with respect to this passage that the basic axiom taken by Mill—the sum of equals is equal—is not even used in the proof with which Mill concludes. In fact, what needs to be used, and is not mentioned explicitly by Mill, is the associative law for addition (Frege, 1884:sec. 9). I would add that Mill's two formulations of his axioms are not equivalent. The second formulation amounts to the transitivity of the "part of" relation. This principle is a basic theorem of Goodman's calculus of individuals, and if Mill's principle that the sum of equals is equal could be shown to be equivalent to it, this would be quite a feat. Indeed, it is very unlikely that even a significant portion of number theory can be derived from the transitivity of the "part of" relation. Only recently has Phillip Sheppard managed to show that a substitute for portions of number theory can be derived within the framework of the calculus of individuals. This result has come after some 30 years of attempts by other able logicians (Sheppard). Mill's carelessness with respect to mathematics certainly was one of the factors that has caused his philosophical view to fall into such disrepute. For it is quite easy to point out mathematical errors in Mill's writings, and when such obvious errors can be found, one suspects that there is something very wrong with the rest of his philosophy. It appears that Mill is not simply making careless mistakes but rather is operating outside his area of competence.

In closing this exposition it should be mentioned that, according to Mill, even arithmetic makes uses of ideal hypotheses, although not in the justification of its definitional postulates. Since these postulates assert the existence of particular numbers and are justified by the existence of particular collections of objects, no use of idealization is needed here. However, among the axioms of arithmetic is the proposition that every number is identical to itself, and Mill thought that at least in some instances this proposition must be treated as an idealized hypothesis. Mill admitted that when we are using arithmetic only in counting populations of apples, idealization is not necessary. But he added that when we use arithmetic in connection with empirical measurements, idealization is essential to avoid empirical falsification of arithmetic. For if we measure two boards and find that one measures 6 feet and the other 8 feet and conclude that the total length obtained by butting the boards will be 14 feet, we are making a series of idealizations. We first idealize when we assume that our boards measure exactly 6 feet and 8 feet, and we next idealize when we assume that the resulting length achieved by juxtaposing the boards is exactly 14 feet. In fact, these assumptions must be idealizations no matter how precise a measuring instrument we use or how many decimal places we use in recording the fraction of feet measured.

Mill's own words are confusing, because he spoke not only about problems of empirical measurement but also about the assumption that the units of measurement do not vary (Mill:Bk. III, chap. VI, sec. 3). This passage led Frege to criticize Mill (Frege 1884:sec. 9), but I think his basic point is well taken. We know from physics that the assumption that a material object such as a table has a precise length is an idealization itself. But even if the concept of the precise length of a table were well defined from a physical point of view, we would still find in practical measurements that a range of numbers can be assigned to the dimensions of the table as it is measured from time to time by different rulers and by different people. So we make use of ranges of error when we measure and add up measurements, and there are well-developed techniques and conventions for handling this problem.

Frege's Criticism of Mill

Frege's criticism of Mill was directed particularly against Mill's analysis of numerical equations and the definitions of individual numbers. It also concerned Mill's claim that numbers are physical properties of conglomerations of physical objects and that the axioms of arithmetic are justified via induction from observations. I shall take up these criticisms in the order just given.

Mill claimed, you will recall, that the definition of 3 as 2 + 1 asserts the empirical fact that certain collections of objects which strike our senses thus $^{O}_{O}{}^{O}$ can be rearranged to strike them thus ∘∘∘. Frege responded to this by asking what would happen to the truth-value of the equation "3 = 2 + 1" if such rearrangements were not possible, and demanded to know the perceptual facts supporting the definitions of large numbers (Frege 1884:sec. 7). Although these remarks only call into question the particular facts or type of facts that Mill cites, there is, I think, implicit in Frege's discussion, a general protest against taking any arithmetical equation as asserting, even in part, a single observable fact. There simply is not a single observation that would bring about the abandonment of our belief that 2 + 1 = 3. If, for example, we combined 2 gallons of milk with 1 gallon of water and obtained less than 3 gallons of watered-down milk, we would not remark upon the refutation of 2 + 1 = 3 but rather would take note of the results of mixing milk and water. To do otherwise would be to make the mistake Frege accuses Mill of making, that is, of confusing "the applications that can be made of an arithmetical proposition, which are often physical and do presuppose observed facts, with the pure mathematical proposition itself" (Frege 1884:sec. 9).

Frege also complained that Mill gave no account of the facts underlying the definitions of zero or one (Frege 1884:sec. 7). Nor is it clear from Mill's approach what those facts could be. For he supports the definitions of individual numbers through facts about the composition and decomposition of aggregates of physical objects. But how do you compose or decompose an aggregate of zero physical objects? There are classes of zero objects but no aggregates or heaps made out of nothing (Frege 1884:sec. 8). Difficulties of this sort led Frege to believe that abstract classes must be viewed as extensions of concepts and dependent upon them.

We might note that although Frege's criticism is quite correct, it would not apply to a more sympathetic reading of Mill. This would construe Mill's account more along the lines of his account of geometrical definitions. Thus, "3 = 2 + 1" would not assert a particular fact but a more general fact that there are aggregates to which the observable property 3 applies which can be decomposed into aggregates to which the observable properties 2 and 1 apply. Then no single contrary experience or even contrary experiences of one or more types need be sufficient to refute an arithmetical law.

Frege remarked that we do not need to make observations to be sure that numerical expressions such as "3 + 1" have a sense (Frege 1884:sec. 8). At that time, he had not introduced his distinction between sense and reference. Mill was concerned with the existence of particular numbers and therefore with the reference of expressions such as "3 + 1." Yet Frege's point is correct whether we interpret it in terms of either sense or reference. Often we are perfectly justified in ascribing both a sense and a reference to a singular term, although we may never have observed the thing to which it purports to refer. Consider the phrases "the animal who left this fossilized footprint" or "the star that was located in such and such a place before it exploded." Now, no one can ever have observed the animal or star in question, and no one may ever be able to carry out such observations. Yet, we are justified in ascribing senses to these expressions on general linguistic grounds, and we might be justified in assigning references to them on the basis of scientific beliefs that are relevant to the particular cases at hand.

Frege denied Mill's claim that arithmetic is about the composition and decomposition of collections by noting that we talk of addition and subtraction, and so on, even with respect to physical collections that we do not literally heap together or break apart. We say ''I have two horses in the back pasture and one horse in the front pasture; I therefore have three horses in toto.'' We make such claims without actually moving the horses around and putting them in one place. For addition is not simply a matter of rearranging and then counting. Furthermore, other things besides physical objects can be counted. We can count ideas, methods, theorems, or numbers themselves. And since there is not even the possibility of heaping these things together, addition cannot be identified with a physical operation. But if addition cannot be identified with a physical operation, then the laws of addition are not laws of nature (Frege 1884:sec. 9).

Frege's argument against Mill's view that numbers are physically perceptible properties of collections of physical objects is one of his most famous and penetrating. It goes as follows. Suppose that I look at two boots, a matched pair. I can say with equal right ''here is a pair of boots, one pair,'' or I can say ''here are two boots.'' In either case what I see is exactly the same, so the same physical fact or physical situation supports both my claims. The difference between the two claims is that I have decided in the one case to count using the term ''pair'' and in the other case to use the term ''boot.'' But if the number of the collections of objects before me were, as Mill claims, a directly observable property of the collection, such as the color of the boots, then the collection would have the number it has without regard to any choice of mine to describe it under one rubric or the other. But, of course, a collection cannot both have the numbers 1 and 2 absolutely (Frege 1884:sec. 25).

Frege had another major objection to all property theories of number, one that applies to Mill but which Frege did not specifically aim at him. The gist of it is this: The grammatical and logical form of numerals and number expressions such as ''the number of stars'' or ''the number ten'' indicates that they are singular terms referring to individuals rather than to properties or concepts. Specifically, these expressions may be substituted for individual variables, may flank the identity sign, and may be used as singular definite descriptions and fail to form plurals (Frege 1884:secs. 38, 51–52, 57). These powerful considerations have been reflected in most analyses of numbers offered since Frege's time. But, of course, a philosopher still has the option of arguing that a deeper analysis than Frege's reveals that number words actually function as predicates. Mill makes a start in this direction, but like most pre-Fregean writers he does not even consider the logical and linguistic evidence which Frege unearthed.

In concluding my discussion of Frege's critique of Mill, I shall briefly present the former's reasons for rejecting the possibility of basing general arithmetical laws upon induction. Inducing a law such as the commutative law of addition from its instances is ruled out by Frege on the grounds that then these instances would need independent justifications and would be deprived of their traditional proofs from general laws and definitions. This consideration does not apply to Mill, however, since he also urged that particular numerical equations are to be deduced from general axioms and definitions. Frege moves on to compare inductions in physics with putative inductions in number theory. His claim is that although most physical properties are invariant under spatial and temporal displacements, the only candidates for projectible properties in number theory are those which refer to the position a number has in the number series, and these are rarely projectible. Thus, induction hardly can find a foothold in number theory. Yet, as Mill noted, there is a crucial property which is invariant in the number sequence: that a number is either zero or can be generated from it by successive additions of one. Frege used the example of a borehole to argue that this does not favor Mill's view. Being a number, he said, is like being one of the core strata from a borehole. These can be characterized as whatever is obtained by starting at the surface and digging deeper. No useful general laws can be formulated in these terms, and the temperatures and rock formations observed at the higher levels cannot be projected reliably to the lower ones. Frege closed by urging that induction itself should be justified via probabilistic considerations and thus would depend upon arithmetic (Frege 1884:sec. 10).

Let us note that Frege's point is not that we cannot reliably project a property that is true of zero and already known to be hereditary in the number series. For he could hardly wish to call into question the principle of mathematical induction. Rather, his point must be that the mere observation that a property is true of all numbers up to a given number is rarely a reliable indication that it holds for numbers beyond that point. Mill's account is too sketchy to determine exactly how and where induction is used to establish the foundations of number theory. Yet we do know that we can verify only a finite number of instances of the number-theoretic axioms, and thus we must—if we follow Mill— project their truth from some point in the number series onto the whole series. And then Frege's point will be applicable.

Is a Millian Account of Number Theory Possible?

Mill's philosophy of arithmetic has many faults, both fundamental and superficial, as Frege has demonstrated successfully. Yet it is not clear that an empiricist or even a Millian philosophy of arithmetic must fail. My aim in this section is to see what faces a more modest program, that of giving a Millian foundation for the natural numbers. I would take such a program to be successful if it could provide an interpretation for the Peano axioms and an analysis of counting in purely physical terms, that is, without appealing to either abstract or mental entities.

Presumably, one would have no serious motivation for embarking on a program of this type unless one were already committed to a concrete physical ontology for other disciplines. (I shall ignore the complication that Mill was committed to a phenomenalistic reduction of material objects.) Thus, we may assume that other purported abstract and mental entities, such as ideas, theorems, and proofs, are capable of a physicalist reduction. This will parry Frege's objection that we can count ideas and theorems as well as apples and oranges.

On the other hand, Mill's view of numbers as properties of aggregates must be abandoned once we have resolved to stick with a physicalist ontology. This can be established by reflecting on that argument of Frege's which I earlier illustrated with the boots example. If classes or other abstract entities are available, then the aggregate of two boots can be distinguished from their pair by identifying the former with the class containing the boots and the latter with the concrete pair itself, that is, with an individual with detached parts. Otherwise, the pair and the aggregate, considered as a concrete individual too, would be indistinguishable and would possess the property of being one as well as that of being two. This difficulty would be obviated if numbers are construed as properties of concrete but disunited individuals consisting of an aggregate and an inscription of a predicate that individuates the parts of the aggregate to be counted. For example, the number 52 could be seen as a property of a deck of cards *plus* an inscription of ''card in this deck.''

There is little to be gained from this, however, bcause concrete numerical judgments such as

> there are nine planets
> the Earth has one moon
> the apples on the table number three
> the number of chairs in the room is six

can be paraphrased as numerical quantifications of the form

> there are exactly n things that are F,

and these in turn paraphrased as sentences in the notation of first-order logic with identity. This avoids treating numbers as properties of individuals and should be welcomed by a Millian, since Mill himself excluded abstract properties from his ontology, despite his account of number (Mill:Bk. I, chap. III, sec. 9).

Moreover, we also can handle concrete comparative judgments such as

> there are more cats than dogs
> there are exactly as many husbands as wives

either by admitting a primitive predicate for each comparison or by admitting some specialized primitive predicates and defining the others in terms of them. For example, by taking ''has more animal parts than'' as primitive ''there are more lions than zebras'' may be paraphrased as ''everything of which every lion and no other animal is a part has more animals as parts than does anything of which every zebra and no other animal is part.'' The reader is referred to Goodman for further discussion.

Without an infinity, or at least a potential infinity of individuals, we have no hope of obtaining the numbers and number theory within this framework. Goodman and his followers have always maintained a neutral position toward infinities, neither admitting nor rejecting them from the nominalist point of view. Other philosophers have suggested supplementing our concrete ontology with modalities so that

> every (concrete) number has a (concrete) successor

can be replaced by

> for every (concrete) number it is possible that it has a (concrete) successor.

Then, for example, suppose that numbers were identified with certain inscriptions of their numerals. Only finitely many numbers would actually exist. But for any given inscribed numeral it is possible to inscribe its successor numeral. Thus, a potential infinity of concrete numbers can be obtained within a modal framework. To keep within the Millian framework it is also essential that we avoid a priori possibilities and stick with empirical and physical ones.

We can also argue for the existence of infinitely many concrete individuals by taking Mill's inductive views more seriously (than they deserve). Let us suppose that events are concrete individuals. On the grounds of simple induction it would follow that every event is followed by another event. Thus, we could identify numbers with some progression of events and establish the Peano axioms inductively. Given the numbers and our predicates for comparative judgments, we then could accumulate inductive evidence for statements of the form

> there are exactly n F just in case there are exactly as many F as there are numbers preceding n.

Furthermore, by defining the number of F via

the number of F = the number n such that there are
exactly as many F as numbers preceding n,

we would obtain

there are exactly n F just in case the number of $F = n$.

This would give us an analysis of counting.

To me, this account is quite strained; and it depends upon the dubious assumption that there are (or it is physically possible for there to be) infinitely many concrete individuals. But it seems this is the direction we must take if we are to satisfy Mill's goal of giving a direct concrete empirical content to the statements of pure and applied number theory. Needless to say, there is little promise of extending this approach to the theory of real numbers. (I have profited from reading unpublished papers on Mill and Millian arithmetic by Glenn Kessler and Philip Kitcher.)

Frege's Philosophy of Mathematics

Many of the goals and presuppositions of Frege's philosophy of mathematics have emerged during our examination of his critical discussions of his contemporaries. From his antipsychologism and antiformalism we may infer that he viewed mathematics as an objective science with genuine truths and a fully meaningful language. His criticism of the early Hilbert demonstrated that Frege also took the assertions of mathematics to be categorical (as opposed to hypothetical) claims, while his attack on Mill dramatized his commitment to the a priori character of mathematics and to a nonphysical interpretation of it. This seems to have left Frege with no alternative except to adopt some version of Platonism, the doctrine that mathematics deals with a domain of abstract entities and studies their properties and relations.

My plan in this chapter is to begin with a discussion of Frege's general views concerning the philosophy of mathematics and work from these to his more specific doctrines concerning the particular mathematical sciences. The exposition will culminate with Frege's analysis of the natural numbers. I will proceed to difficulties with Frege's views that have been raised by subsequent developments, and conclude with an evaluation of Frege's philosophy of mathematics and its impact.

General Considerations

Frege seems to have had no alternative but to adopt Platonism, but two thoughtful and learned interpreters, Hans Sluga and Michael Dummett, have engaged in a heated controversy concerning Frege's realism

(Sluga 1975; 1976; 1977; Dummet 1976). We have little hope of disentangling this unless we specify the terms "realist" and "Platonist" more carefully than has been done by the parties to the dispute. Let me start by introducing the term *methodological Platonist* to characterize someone who endorses nonconstructive mathematical methods such as the use of excluded middle, impredicative definitions, sets, and the like. Such mathematicians have been called Platonists because there is at least a *prima facie* case that the use of nonconstructive methods presupposes a picture of mathematics as dealing with a mind-independent infinite domain of abstract entities. No one has any doubts about Frege's position in this regard. He made full use of the excluded middle, sets, and impredicative definitions.

Let us call an *ontological Platonist* someone who recognizes the existence of numbers, sets, and the like as being on a par with ordinary objects and who does not attempt to reduce them to physical or subjective mental entities. An *epistemological Platonist* is someone who also believes that our knowledge of mathematical objects is at least in part based upon a direct acquaintance with them, which is analogous to our perception of physical objects. Finally, a *realist* believes that the objects of mathematics (and empirical science) exist independently of us and our mental lives. The realist is opposed not only to subjective idealists but also to transcendental or objective idealists such as Kant. Thus, a nonrealist could endorse all three forms of Platonism while arguing that a deeper analysis of mathematical "perception" shows that the mind or reason constructs the whole world of mathematical objects prior to our "experiencing" it. The effect of this would be to make such a world *objective* in the sense of being "independent of our sensation, intuition, and imagination, and of all construction of mental pictures out of memories of earlier sensations . . . " (Frege 1884:sec. 26) but still not a world of things in themselves. If I am correct in this claim, then the distinction between a Platonic realist and a Platonic objective idealist is not a crucial one for current philosophy of mathematics. For today the focus is upon the methodology and semantics for mathematics, and the two forms of Platonism are likely to lead to the same conclusion in this area.

It is certainly important in coming to understand Frege, however, to

[*162*]

decide where he fits in the scheme just introduced. Sluga associates Frege with objective idealism, rationalism (which, fortunately, we need not define), and the philosophical tradition of Leibniz, Kant, and Lotze. Sluga relies heavily upon historical arguments, Frege's distinction between objectivity (*Objectivität*) and reality (*Wirklichkeit*), and his context principle (i.e., the principle that words have meanings only in the context of a sentence). Dummett has countered by downplaying the historical connections, pointing out that the translation of *wirklich* as "real" or "actual" is misleading in the context of a dispute over Frege's realism and arguing that Frege abandoned his context principle at the time he evolved his theory of sense and reference. (Ironically, I argued the last two points and criticized Dummett for attributing the context principle to the post-1891 Frege [Resnik 1967]. I do not believe that Dummett knew of this article.)

My own view is that Sluga's interpretation is an extremely plausible and well-supported view of Frege's *Grundlagen*. I shall argue also that the *Grundlagen* is a difficult work to interpret because Frege is not entirely consistent in his philosophical pronouncements. On the other hand, I must side with Dummett concerning Frege's abandonment of the context principle and his subsequent adoption of what appears to be a thoroughly realist stance. I do not agree with Dummett's claim that Frege *should* have retained his context principle, but this is due largely to my failure to understand Dummett's account of how the context principle would fit with Frege's later doctrines.

In the *Grundlagen* there are many remarks that hint that numbers are not completely independent of the mind. Frege says that they are objective, but in the same section he distinguishes the objective from the real (Frege 1884:sec. 26). This is also the section in which he asserts both the objectivity of geometry and its intuitive foundation. This assertion is followed by the long passage, quoted in Chapter One, containing the example of the geometry of two beings whose spatial intuitions are duals. The section concludes with the remark that the objective is

independent of our sensation, intuition, and imagination, and of all constructions of mental pictures out of memories of earlier sensations, but not what is independent of the reason; for to

[*163*]

undertake to say what things are like independent of the reason, would be as much to judge without judging, or to wash the fur without wetting it. [Frege 1884:sec. 26]

In a footnote to the next section, Frege writes that he has no objection to *objective ideas* that "are the same for all" and "can be divided into objects and concepts" (Frege 1884:sec. 27). After arguing at length that numbers are self-subsistent objects, he qualifies his doctrine with the remark:

> The self-subsistence which I am claiming for numbers is not to be taken to mean that a number word signifies something when removed from the context of a proposition, but only to preclude the use of such words as predicates or attributes, which appreciably alters their meaning. [Frege 1884:sec. 60]

All of this leads the reader to believe that Frege's numbers are not full-fledged entities, that the objectivity of arithmetic—like that of geometry—does not result from the existence of a mind-independent domain of numbers which can be apprehended by many, and that numerals cannot be assigned their references through baptism or ostension or any process analogous to that through which, say, proper names of persons acquire reference. In addition, the *Grundlagen* contains many passages praising Kant, whose basic epistemological framework Frege adopted (Frege 1884:secs. 3, 89). If we add to this the data Sluga has unearthed concerning Frege's exposure to Lotze's ideas and the similarity of Frege's distinction of the objective vs. the real to Lotze's distinction of the objective or valid vs. the real or existent (Sein), then the case for Frege's transcendental idealism is most persuasive.

Indeed, one might also argue that Frege was merely a methodological Platonist during the *Grundlagen* period. For in establishing how numbers are "to be given to us" he states that, in view of the context principle, we may obtain them by explaining the sense of sentences in which number words occur (Frege 1884:sec. 62). Thus, it could be argued that Frege aimed at giving a nonreferential truth-value semantics for arithmetic, that is, a semantics in which sentences receive truth-values directly and the "references" of their parts are construed as abstractions or a *façon de parler*. (Cf. Frege 1969:18, note 2.) For example, on this view a proper name would be construed as referential provided that every sentence in which it occurs has a truth-value (as long as the other parts were referential), and two proper names would have the same reference provided that the identity between them is true.

Unfortunately for his interpreters, Frege's other pronouncements during the *Grundlagen* period cloud this interpretation of him as a non-Platonic transcendental idealist. First, there are comparisons of numbers with real (*wirklich*) objects such as the earth (Frege 1884:sec. 60). Second, Frege comes close to allowing that numbers are *wirklich* in this passage concerning transfinite numbers:

> I heartily share [Cantor's] contempt for the view that in principle only finite numbers ought to be admitted as *wirklich*. Perceptible by the senses these are not, nor are they spatial. . . . and if we restrict the *wirklich* to what affects our senses or at least produces effects which may cause sense-perceptions as near or remote consequences, then naturally no number of any of these kinds is *wirklich*. [Frege 1884:sec. 85. (Austin has here translated "wirklich" as "existent.")]

Third, in a piece written around 1884, Frege argues that sentences of the form "$(\exists x)\ (x = N)$," where N is a name, are trivially true since logic presupposes that "the words are not empty, that sentences are expressions of judgments, that we are not merely playing with words" (Frege 1969:67). It seems implicit in this passage that not only are we not free to stipulate truth-values for sentences but also that their having truth-values presupposes that the names occurring in them have references. (This presupposition doctrine became explicit in "On Sense and Reference.")

I know of no way to eradicate all the tension in Frege's views which these contrary themes produce, but I think that the tension can be eased considerably through another interpretation of the role of the context principle in introducing numbers. Consider this passage:

In arithmetic we are not concerned with objects which we come
to know as something alien from without through the medium of
the senses, but with objects given directly to our reason and, as
its nearest kin utterly transparent to it. [Frege 1884:sec. 105]

Frege raised the question "How are numbers given to us?" and answered it via the context principle by directing us to sentences containing number words (Frege 1884:sec. 62). I propose the following interpretation of his answer: Numbers are given directly to our reason. But
the elements of reason are judgments, since we do not reason with
words but rather with sentences and in so doing make judgments. Thus,
to determine the nature of numbers we must look at judgments of
number and about numbers. Do not ask what number words mean
outside the context of a sentence. That is like asking what numbers are
like independently of judgments, which in turn is to ask what they are
like independently of reason. And to answer that would be to "judge
without judging."

The virtue of this reading of Frege is that it retains his ontological
platonism while explaining the importance of the context principle to his
analysis of number. The truth-value semantics account of the principle
is not that attractive, because in fact no such semantics is given for
number words in the *Grundlagen*. They are given explicit definitions in
terms of extensions of concepts. The proposed reading also shows the
important role of rationalist themes in Frege's thought. Thus, although
no interpretation can be confirmed definitely or refuted by Frege's own
words, the one that is best supported and ties together the most strands
in the *Grundlagen* is the one I have sketched. To summarize: In the
Grundlagen, Frege was an ontological Platonist and an objective
idealist. The application of the context principle to the analysis of
number is a move within the rationalist tradition which seeks to show
how our knowledge of arithmetic is based upon the faculty of reason.
(In Resnik (1976), I also discuss the role of the context principle in
Frege's philosophy of language.)

The context principle is one of the three "fundamental principles" of
the *Grundlagen*. The other two promulgate the distinction between the
subjective and the objective and the distinction between concepts and

objects (Frege 1884:x). The latter principles are advanced and defended repeatedly in Frege's later works, while the context principle itself is never presented again. In the *Grundlagen* the context principle was used against psychologism—it asks for meanings of words in isolation from sentences—but this accusation or attack is never used again in Frege's subsequent refutations of psychologism. Nor are we told again that numbers are given directly to reason. Furthermore, there are passages and themes in Frege's later works which explicitly contradict the context principle: (1) After 1891 no logical or semantic distinction is made between sentences and proper names; both are used to refer to objects (Frege 1891:13–14). (2) In later works Frege held that we can understand sentences we never have heard before because "parts can be distinguished in the thought which correspond to sentence parts, so that the construction of the sentence can be a model of the construction of the thought" (Frege 1923:36). (3) In a letter to Peano, Frege wrote:

> The task of our natural language is essentially fulfilled when the people that communicate in it connect the same thought with the same sentence, or, at least, approximately the same. For that it is completely unnecessary that the words taken by themselves have a sense and reference as long as the whole has a sense. The matter is different when inferences are supposed to be drawn. Then it is essential that in two sentences the same expression occurs and that in both it has exactly the same reference. The expression must therefore have a reference that is independent of the other parts of the sentence. [Frege 1896:55–56]

Those who wish to argue that Frege retained the context principle can find *traces* of it in Frege's later works. Sluga even argues that the passage just quoted supports the principle:

> What this passage says can best be stated by distinguishing three epistemic stages. At the first, that of ordinary language, senses of sentences can be apprehended without prior apprehension of the senses and references of the parts of sentences. [Sluga 1977:240]

Sluga overlooks the fact that Frege makes these remarks in connection with his doctrine that a predicate must be regarded *from the logical point of view* as without reference if there are objects to which it is undetermined linguistically whether the predicate applies. Thus, a more fitting reading of Frege would be that some words of ordinary language have faulty meanings whose unclarity usually gets resolved by the sentential context containing them. The reason Frege gives for requiring independent references in inference is that it avoids equivocation—a word's being used in two sentences with different references. But equivocal words do not lack meaning in isolation. Rather, they have too many meanings in isolation. In any case, Frege's requirement that in an ideal language words have meaning independent of context surely contradicts the context principle, since Frege never indicated that it was restricted to ordinary language.

A more formal trace of the context principle can be found in section twenty-nine of the *Grundgesetze,* where Frege lays down conditions for names to have a reference. Here he says that a proper name has a reference provided that a referential name is produced whenever it is substituted for an argument place-holder of a referential function name. This is quite close to the view that a proper name is referential provided that every sentence composed of it and referential names has a truth-value. But it must be noted that Frege explicitly states that these conditions are not definitions "because their application always presupposes that we have already recognized some names as denoting" (Frege 1893:sec. 30). These conditions are in fact crucial premises in Frege's attempted inductive proof that all correctly formed names of his symbolism have a reference. (This proof has a number of faults. One is that a unary function name may give rise to a referential name whenever a referential object name is put for its argument place and yet the "function" it denotes may not be everywhere defined. See Resnik (1963) for a detailed discussion of Frege's proof.)

Next let us look at Frege's use of the term *wirklich* in his later writings. In the *Grundgesetze* he distinguished existence (which is a property of concepts) from objectivity and actuality (*Wirklichkeit*) (which are properties of objects) (Frege 1893:xxv). Here he also explains the actual as "capable of acting directly or indirectly on the senses,'' and urges that we should recognize the existence of numbers and thoughts and concepts even though they are not actual (Frege 1893: XVII). Thus, the contrast between the actual and nonactual is being used by him as the contrast between the physical and nonphysical rather than as the contrast between the real and the ideal. Furthermore, in a later essay Frege even says that thoughts are *wirklich* in an extended sense:

> The thought, admitted is not something which it is usual to call real (*wirklich*). The world of the real is a world in which this acts on that, changes it and again experiences reactions itself and is changed by them. . . .
>
> And yet! What value could there be for us in the eternally unchangeable which could neither undergo effects nor have effects on us? Something entirely and in every respect inactive would be unreal and non-existent (*nicht vorhanden*) for us. . . . But by apprehending a thought I could come into a relation to it and it to me. . . .
>
> How does a thought act? By being apprehended and taken to be true. . . . Thoughts are by no means unreal but their reality is of quite a different kind from that of things. [Frege 1918:76-77. This passage is also quoted in Currie, where it is argued that the later Frege was a realist.]

In Frege's later writings the *mind-independence* of thoughts and numbers and other abstract objects receives a strong emphasis that is not found in the *Grundlagen*. Frege speaks of a realm or domain which is objective and yet not sense-perceptible whose members are *already on hand* to be "grasped" or "apprehended" by different people (Frege 1893:XVIII, XXIV; 1918:69-70, 74-75). I fail to see how these metaphors can be reconciled with the nonrealism evident in the following passage from Lotze, which Sluga has quoted as a precursor of Frege's idealism:

> Through the logical objectification, which is indicated by the creation of a name, the identified content is not put into an external reality; the common world in which others are to find

this content to which we are pointing is, in general only the world of the thinkable; to it we ascribe only the first trace of an independent subsistence and of an inner regularity, which is the same for all thinking beings and which is independent of them. And it is, therefore, of no concern whether any part of this world of thought refers to anything which has an independent reality apart from thinking minds, or whether its whole content persists only with equal validity in the thinking of the thinkers. [Lotze:16; Sluga 1977:223]

I agree that this fits nicely with the *Grundlagen*. There Frege even characterized logic and arithmetic as governing "everything thinkable" (Frege 1884:sec. 14). Yet, it also seems to me that we get a much clearer and simpler interpretation of Frege if we grant that he changed his mind and admitted an abstract world that exists independently of all thinking and thinkers.

To finish the case for my view, I appeal to a piece of evidence that neither Sluga nor Dummett has discussed: the appearance in 1891 of the metaphor of *unsaturatedness* in Frege's doctrine of incomplete entities. This metaphor has always been a weak point in Frege's theory. He recognized the problems it leads to rather early (Frege 1892b), and later anticipated the problems for the very formulation of his semantics that it produces. (Cf. Resnik 1965.) Why did he cling to it so steadfastly? The answer is to be found in these passages:

The whole [thought] owes its unity to the fact that the thought fulfills the unfulfilled part or, as we can say, completes the part needing completion. And it is natural to suppose that, for logic in general, combination into a whole always comes about through the fulfillment of something unfulfilled. [Frege 1923:37]

An object—e.g. the number 2—cannot logically adhere to another object—e.g. Julius Caesar—without a binding agent. But that may not be an object, but must be something unsaturated. A logical combination can only become a whole by having an unsaturated part saturated or supplemented by one or more parts. [Frege 1903b:1, 372]

But why would Frege need this logical cement if he still had the context principle? I agree with Sluga that Frege was deeply concerned with the Kantian problem of the unity of thought or judgment (Sluga 1977:238) and that the context principle was a solution to it. But this solution was superseded in 1891 by the theory of unsaturated senses and references.

In Frege's later works epistemological Platonism is also evident:

> We do not have a thought as we have, say, a sense-impression, but we also do not see a thought as we see, say, a star. So it is advisable to choose a special expression and the word "apprehend" offers itself for the purpose. A particular mental capacity, the power of thought, must correspond to the apprehension of thoughts. In thinking we do not produce thoughts but we apprehend them. [Frege 1918:74]

Frege never explains how we manage to apprehend thoughts or other abstract entities. In an unpublished work of 1897 he grants that "the process is the most mysterious of all" (Frege 1969:157). One wonders whether on Frege's view there is any reason to expect this mystery to be explicable. For if we have no access to someone else's inner world, how will we ever arrive at a scientific treatment of our knowledge of abstract entities? We can also observe the star that someone else sees; we can observe his eyes, retinas, and brain and thus develop a scientific account of perception. But how, for example, can we distinguish between a veridical and an illusory apprehension of a thought?

Axioms, Definitions, Inference, and the Epistemology of Mathematics

Thus far, I have argued that in his mature writings Frege espoused both an epistemological and ontological Platonist philosophy of mathematics. Let us now explore his epistemology somewhat further. Frege recognized Kant's fourfold classification of knowledge, although he defined it somewhat differently. On Frege's view, a truth is a priori if and only if its proof rests upon general laws "which neither need nor admit of proof" and does not appeal to facts (i.e., "truths which cannot be proved and are about particular objects"). He defines a truth as analytic if and only if its proof depends only on general logical laws and definitions, including any proposition on which the correctness of the definitions may depend (Frege 1884:sec. 4). This last provision probably has to do with definitions (such as those using definite descriptions) that presuppose uniqueness and existence theorems.

Frege's characterization of the epistemological character of a truth in terms of its proof raises a serious difficulty that he himself hinted at in the following passage:

> We very soon come to propositions which cannot be proved so long as we do not succeed in analyzing concepts which occur in them into simpler concepts or in reducing them to something of greater generality. [Frege 1884:sec. 4]

The difficulty is this: How do we know when we have the epistemologically ultimate proof of a truth? If a truth has two very different proofs, how do we decide which one is the epistemologically ultimate one?

To make the matter clearer, let us consider some cases. First take a truth of number theory, for example, the associative law of addition. This can be proved from the Peano axioms, but it can be derived also in the theory of finite sets. Since the latter proof depends upon analyzing number-theoretic concepts in terms of set-theoretic ones, it would be safe to conclude that from Frege's viewpoint it is the more ultimate proof. But can we conclude that it is *the ultimate proof*?

Next consider the case of real number theory and geometry. When Frege wrote the *Grundlagen,* he knew that geometry can be modeled within the theory of real numbers. His early mathematical work was even in analytical geometry (Frege 1873)! Furthermore, the reduction of geometry to real number theory is of the same formal type as is his reduction of arithmetic (including real number theory) to set theory. Why, then, did he adamantly claim that geometry was synthetic a priori? What made a proof of a geometric truth from geometric axioms more ultimate than one from the axioms of set theory?

Frege's writings contain several passages asserting that logical truths are more general and basic than geometrical or other synthetic a priori truths. Geometrical truths can be defined without destroying the possibility of thought, but when we try denying a logical or arithmetical truth "even to think at all seems no longer possible" (Frege 1884:sec. 14; cf. Frege 1885:95–96). Yet Frege did not appeal to this to resolve our difficulty. Furthermore, given our current knowledge of mathematics, no resolution in terms of generality is possible. For we now know of several general foundations for mathematics—set theory and category theory—which can be reduced to each other, and there are no clear mathematical grounds for giving priority to any of these "foundations." I shall argue below, however, that this is not crucial to Frege's work, since his aim was the rather different one of determining the place each mathematical theory has in his version of the Kantian framework.

As Frege tied the epistemology of mathematics to the axiomatic method, it is important for us to examine his views on definitions and axioms. We have discussed the methodological distinction between axioms and definitions in connection with the Frege-Hilbert controversy and noted that for Frege axioms are not simply the starting points of mathematical proofs. To be sure, he recognized that a statement's being an axiom depends upon its role in the particular systematic development of the branch of mathematics to which it belongs. He also knew that different axiom systems are possible for the same branch of mathematics, so that the property of being an axiom is not an absolute one but is relative to a system (Frege 1969:221–222). Despite this, Frege required that each axiom satisfy two nonsystematic conditions: it must be true and there must be no doubt as to its truth. His reasons were: (1) if there is any doubt as to the truth of a proposed axiom, then it should be proved rather than accepted without proof; and (2) one certainly should not state something as an axiom which he knows to be false (Frege 1969:221).

The modern mathematician certainly would object to Frege's view of axioms by pointing out that we sometimes set up axiom systems for the purposes of a purely deductive investigation without considering whether they are true. From the logical point of view this is the same as the deductive investigation of the consequences of a premise that is either known to be false or whose truth is at least doubtful. Frege had considered this objection and answered that we do not and cannot make inferences from statements whose truth value is doubtful or known to be false (Frege 1969:264; 1906a:424). On Frege's view, inferences are not transformations on sentences; they are psychological acts that can be linguistically manifested but need not be.

> An inference does not belong to the domain of signs, but is rather a matter of making a judgment which is based upon previously made judgments in accordance with logical laws. Each premise is a definite thought which has been recognized as true, and the conclusion is also a definite thought which is recognized as true. [Frege 1906a:387]

Furthermore, conditional and indirect proofs are, on Frege's view, only apparent exceptions to this claim.

> The assumptions [of a conditional proof] are not premises of inferences. The premises are rather some hypothetical thoughts which contain the thoughts in question as conditions. These also appear in the end result; and it follows from this that they are not used as premises since if they were they would disappear from the end result. If someone omits them he has made a mistake. Only after the thoughts in question have been recognized as true may one drop them as conditions. This is done via an inference whose premises are the thoughts that are now recognized as true. [Frege 1906a:425]

Frege even analyzed some examples to show that the use of conditional and indirect proofs can be eliminated from them (Frege 1969:264–266). His treatment is anticipatory of the Deduction Theorem but is far too specific to earn him credit for having formulated the theorem or for indicating its proof.

Frege's odd pronouncements about axioms are immediate consequences of his theory of inference. For every mathematical proof has axioms as its ultimate premises. Hence, they must be clearly recognized as both true and not needing proof. As this and Frege's reductionistic epistemology place a heavy burden upon his notion of truths which "neither need nor admit of proof," it is a pity that he never explained that crucial phase.

Frege also made frequent use of the term "self-evident" and distinguished between logical and intuitive (*anschaulich*) self-evidence (Frege 1884:sec. 90). Clearly, there is a connection between this notion and the phrase "neither need nor admit of proof." A self-evident statement needs no proof—since there is no doubt as to its truth—nor will it be capable of a proof that yields additional conviction or justification. In this connection Frege remarked that his axiom V, the one that introduces classes, lacked the self-evidence that must be required of a logical law (Frege 1903a:253). However, Frege never published an account of how axioms are to be justified, and in the following passage seems to forswear the task:

> The question why and with what right we acknowledge a law of logic to be true, logic can only answer by reducing it to another law of logic. Where that is not possible, logic can give no answer. If we step away from logic, we may say: we are compelled to make judgments by our own nature and by external circumstances; and if we do so, we cannot reject this law—of identity, for example; we must acknowledge it unless we wish to reduce our thought to confusion and finally renounce all judgment whatever. *I shall neither dispute nor support this view;* I shall merely remark that what we have here is not a logical consequence. What is given is not a reason for something's being true but for *our taking it to be true.* [Frege 1893: XVII; my emphasis]

Indeed, as this passage occurs in a discussion of psychologism's conflation of truth with acceptance as true, many commentators have concluded that Frege was not interested in epistemological questions except insofar as they can be formulated as questions about mathematical proof. There is no doubt that these are the only epistemological questions upon which Frege worked, but there is also evidence that he placed them in a broader and essentially Kantian framework.

In the *Grundlagen* Frege declares that epistemological questions partially motivated his investigations of the foundations of arithmetic in that he wished to determine how it fits within the analytic-synthetic, a priori–a posteriori matrix (Frege 1884:sec. 3). It is also here that he praised Kant generally—and specifically for maintaining that geometry is synthetic and a priori (Frege 1884:sec. 89). Nor is this agreement with Kant restricted to Frege's earlier period; throughout his life he maintained that Euclidean geometry is true and based upon intuition (Frege 1903b:319; 1912:240–241; 1941:14; 1969:182–184).

This stance vis-à-vis Euclidean geometry comes through as quite dogmatic in an unpublished paper written between 1899 and 1906. Frege begins by explaining how science is a system of truths and why he requires that axioms be true. He then continues as follows:

> If Euclidean geometry is true, then non-Euclidean geometry is false, and if non-Euclidean geometry is true then Euclidean geometry is false. . . .
>
> We once took ourselves to be pursuing a science which we named Alchemy; but when we recognized that this alleged science was thoroughly erroneous, we banned it from the ranks of science. . . . Now the question is whether to strike Euclidean or non-Euclidean geometry from the ranks of science and to put it alongside of Alchemy and Astrology as mummies. Where one only let himself toy with ideas, he need not take things so earnestly, but in science the quest for truth must impose a strong dominion. That means: either in or out! Now is Euclidean or non-Euclidean geometry supposed to be flung out? That is the question. Does one dare to treat Euclid's Elements, which have maintained an undisputed authority for over 2000 years, like Astrology? If one does not dare to, then one can only set forth Euclid's axioms as neither false nor dubious. Then non-Euclidean geometry must not be counted as scientific and assessed as only an historical oddity of little importance. [Frege 1969:183–184]

Aside from being surprised at this harsh rejection of non-Euclidean geometry (which may have been justified by the concurrent state of physics), the reader is probably wondering why Frege could not allow other geometries to describe mathematical structures different from Euclidean space. Perhaps he could, but Frege posed the question so that we are to decide whether the parallel postulate qua thought or sentence meaning—not qua uninterpreted sentence—is true or false. He simply leaves no room for hedging with "it's true in Euclidean space, false otherwise." Indeed, this position led him to reject Hilbert's independence proofs:

> The word "interpretation" is objectionable; since a thought, correctly expressed, leaves no room for different interpretations. [Frege 1906a:384]

> ... but Mr. Hilbert's axioms are improper sentences which thus express no thoughts. We see this from the fact that according to Mr. Hilbert an axiom sometimes holds and sometimes does not. A proper sentence expresses a thought and this is true or false. [Frege 1906a:424]

Yet I do not think that Frege's position on geometry was simply a reaction to Hilbert. In a manuscript of 1924 or 1925 called "The sources of knowledge of mathematics and the mathematical natural sciences," Frege introduces three sources of knowledge: sense experience, the logical source, and the geometrical-temporal source. Most of his discussion focuses upon the reliability of each source. Sense experience sometimes produces sensory illusions, and the close tie between thought and language is responsible for both logical illusions in the form of improperly referring singular terms—such as "the concept *horse*"—and the paradoxes of set theory. The geotemporal source, from which the axioms of geometry and our knowledge of infinity comes, is the most reliable source of knowledge (Frege 1969:286–294). Unfortunately, Frege does not explain how these sources produce knowledge or why they do. Philip Kitcher has proposed that Frege believed that Kant had already answered these questions (Kitcher 1979). Be that as it may, the analogy to Kant's three types of knowledge is striking; and, although the essay on the three sources was written in Frege's last years, the Kantian epistemological framework dates from his earliest works.

The last essay, moreover, begins to make sense of several of the otherwise puzzling features of Frege's doctrines: on the one hand, his commitment to Euclidean geometry flows naturally from his belief in the reliability of geometrical knowledge, while on the other hand, self-evident truths are those which arise directly from an a priori knowledge source. It seems to me that Frege saw his own task as that of determining to which source each of the various branches of mathematics reduces. The basic method for carrying out this program consisted of first determining the axiomatic basis for each branch and then the knowledge source from which those axioms flow: "The problem becomes, in fact, that of finding the proof of a proposition, and of following it up right back to primitive truths" (Frege 1884:sec. 3).

But if this is what Frege's epistemological views come to, then concern about the ultimate proof of a truth seems even more pressing. If geometry can be "reduced" to arithmetic and this in turn to logic—all of which Frege should have granted—then why are there *three* knowledge sources? Why is geometry nevertheless synthetic a priori for Frege? The answer is found, I think, in Frege's nontechnical discussions of geometry and arithmetic. Geometry, he tells us, governs only our spatial intuitions, whether real or imagined; its axioms can be denied without contradiction for the purposes of "conceptual thought" (Frege 1884:sec. 14). Furthermore, since one spatial figure can represent any other to which it is congruent and cannot be distinguished from them except when several are presented together, "it is quite intelligible that general propositions should be derived from intuition; the points or lines or planes which we intuit are not really particular at all, which is what enables them to stand as representatives of the whole of their kind" (Frege 1884:sec. 13).

Arithmetic, on the other hand, is far more general; one cannot deny its laws even for the purposes of conceptual thought. Since everything is numerable, the laws of arithmetic and the laws of logic govern the same domain (Frege 1884:sec. 14). Furthermore, intuition seems an unlikely foundation for arithmetic because

> even number has its own peculiarities. To what extent a given
> particular number can represent all the others, and at what point

[*178*]

its own special character comes into play, cannot be laid down generally in advance. [Frege 1884:sec. 13]

(Frege gave several other reasons why arithmetic cannot be based upon intuition, but these also tell against an intuitive basis for geometry. For example, he argued that we cannot intuit large numbers (Frege 1884:sec. 5), but one can raise the same objection with respect to say a 10^{10}-sided polygon.)

Frege's informal epistemological discussions of arithmetic and geometry should be regarded as essential parts of his argument rather than as mere motivating considerations for formal developments. For by introducing epistemological constraints, they serve to rule out formally correct but epistemologically unsatisfactory reductions of arithmetic to geometry or of geometry to set theory. Frege did not consider explicitly the issue of alternative reductions, but his rejection of geometrical foundations for arithmetic and Newton's account of number are based upon nontechnical considerations. In discussing Newton's definition of a number as a relationship between magnitudes, Frege raised the question as to whether it presupposes a definition of cardinal numbers but *left it in favor of more philosophical considerations:*

> However, let it be, as it may be, the case that identity of ratios between lengths can in fact be defined without reference to the concept of number. Even so, we should still remain in doubt as to how the number defined geometrically is in this way related to the number of ordinary life which would then be entirely cut off from science. Yet surely we are entitled to demand of arithmetic that its numbers should be adapted for use in every application made of number, even though that application is not itself the business of arithmetic. [Frege 1884:sec. 19]

The foregoing passages and considerations lead me to take a view of Frege's dispute with Kant concerning the analyticity of arithmetic which differs from that suggested by Frege himself in his motivational discussions of his formal work. (Cf. the introduction to 1879b and 1893.) There he seems to hold that the existence of a formal derivation of arithmetic from logic alone would refute Kant. I propose, however, that if pressed he would admit that this is only a necessary condition for the establishment of the thesis that arithmetic is analytic. The formal reduction must be supplemented with an epistemological analysis that shows the proper place of arithmetic in Frege's modified Kantian framework.

According to Frege, the main modification that he made in Kant's epistemological schema consisted in the introduction of more powerful forms of definitions for use in formal proofs (Frege 1884:sec. 88). Now any statement can be proved from "definitions" if we countenance certain forms of definition. For instance, the theorems of geometry would be derivable from definitions if implicit definitions via axioms were recognized as true definitions. Thus, in evaluating Frege's thesis that arithmetic is part of logic and thus analytic, we must pay careful attention to his theory of definitions. Frege sees definitions as assigning sense and reference to the symbols they introduce by postulating that the *definiens* is to have the same sense and reference as the *definiendum*. From a logical point of view, then, definitions are abbreviatory conventions which allow us to prove new sentences but not new thoughts or propositions (Frege 1879b:sec. 24). Frege did not explicitly formulate the criteria of eliminability and noncreativity for definitions. This was done subsequently by Lesniewski (Suppes). The criteria are clearly implicit in Frege's writings, however, and he did give rules of definitions which guarantee that definitions introduced according to them satisfy the two criteria (Frege 1893:sec. 33).

Frege gave his definitions in the object language. In his semantics he treated sentences as names of truth values. Then "$A \equiv B$," which is true if and only if A and B have the same truth-value, can be written "$A = B$." This fuses identity and the biconditional with the result that every definition is formulated as an identity. This, in turn, led Frege to hold that identity must be taken as primitive and cannot be defined (Frege 1894:320–321). Frege wrote definitions in the following form: $\Vdash A = B$. Once a definition has been introduced into the system, it may be rewritten as an assertion as follows: $\vdash A = B$. It can then be used as a premise in proofs as if it were an axiom (Frege 1893:sec. 27; 1969: 224–225). This approach to definitions allows him to dispense with a separate definitional replacement rule and simply make use of the substitutivity of identity.

Frege recognized that it is crucial to his view of arithmetic as analytic that definitions not introduce synthetic knowledge. He defended this by arguing that when a definition is first given, one does not introduce a truth at all, simply a meaning stipulation. Only after a definition has been introduced and the symbol presented through it has been given a meaning can one raise the question of its truth. Yet, one then sees that since both its sides have the same sense and reference, it is simply an instance of the law of identity $a = a$ and analytic (Frege 1879b:sec. 24; 1903b:320). Although definitions do not introduce new knowledge, Frege claimed that they are indispensable and provide valuable insights into logical connections between truths. He even declared that those definitions whose omission leaves no gaps in proofs should be rejected as worthless (Frege 1884:sec. 70; 1906a:302–303). How are we to reconcile these two themes? And how shall Frege resolve the problem of analysis posed by Husserl's critique (Frege 1894:319)?

I think that these questions prompted a certain amount of tension in Frege's thinking and eventually led to a change in his views. The *Grundlagen* presents the definition of number as the result of an analysis of the concept of number. Frege assumes that those who follow his analysis will see that his definitions only make explicit what was already implicit in our concept of number. Each definition can thus be judged according to the accuracy of the analysis it reports.

In responding to Husserl's criticisms of this definition, Frege makes the first qualification in this view: As far as mathematics is concerned, definitions must preserve only extensional equivalence (Frege 1894:319–320). But as noted in our discussion of Frege and Husserl, this response is not adequate. By Frege's time, mathematicians already knew of alternative definitions of the real numbers in terms of sets that are not even extensionally equivalent.

In 1906, Frege took a further step toward coping with these problems by distinguishing between the role that definitions play in an axiomatic system and the conceptual activity that leads to them: The following passage presents his view well:

The mental activities leading to the formulation of a definition may be of two kinds: analytic or synthetic. This is similar to the activities of the chemist, who either analyzes a given substance into its elements or lets given elements combine to form a new substance. In both cases, we come to know the composition of a substance. So here, too, we can achieve something new through a logical construction and can stipulate or sign for it.

But the mental work preceding the formulation of a definition does not appear in the systematic structure of mathematics; only its result, the definition, does. Thus, it is all the same for the system of mathematics, whether the preceding activity was of an analytic or a synthetic kind; whether the definiendum had already somehow been given before, or whether it was newly derived. For in the system, no sign (word) appears prior to the definition that introduces it. Therefore so far as the system is concerned, every definition is the giving of a name, regardless of the manner in which we arrived at it. [Frege 1906a:303]

The implication of this passage, then, is that the epistemological status of a truth is to be judged by reference to the axiom system containing it. If it can be derived from logical laws and definitions alone, then it counts as analytic. This holds regardless of whether the definitions used introduce new symbols and concepts or have resulted from an analysis of previously given symbols or concepts. Now both "analytic" and "synthetic" definitions can produce important deductive consequences within an axiomatic system, although only the former can lead to new proofs of an already accepted truth. Frege would strongly resist basing any conclusions concerning the analyticity of a statement upon this distinction between definitions. For it is an historically based distinction, not a logically based one; and to import it into his epistemology would risk an involvement with psychologism.

This attempt to reconcile the epistemic triviality of individual definitions with their systematic fruitfulness goes a long way, but it does not respond to the problem of the criteria for the correctness of individual analytical definitions. In an essay of 1914, Frege made the final change in this theory in order to deal with this problem. He began by reiterating his previous views concerning analytical activity and the formal role of definitions. This time he also emphasized the pragmatic necessity for definitions. They allow us to abbreviate proofs and make them surveyable. Whatever logical analysis may have been used in obtaining the definitions is not reflected in the system. As far as the system is concerned, there is no distinction between "classes of classes in one-to-one correlation with some class" and "number." But we could not operate solely with the former phrase. Even when we use the term "number" in the defined sense, we are not consciously aware of its expanded meaning. Only when we need to refer back to its full meaning in the course of constructing a proof do we rely upon the definition. Moreover, it would be impossible for us to operate with the system if we had to keep in mind the full meaning of each of its symbols.

Turning next to the problem of the correctness of a definition, Frege distinguished two cases. In the first case, the definition introduces a brand new symbol, as when we define "R" as "$\hat{x}(x \notin x)$." Here the definition is a purely abbreviatory one, and the question of presystematic meaning does not arise. In the second case, the symbol being defined is already in use, as in Frege's definition of "number." Here Frege recognized two subcases. In the first subcase the symbol already has a clear meaning. Then the definition is correct if and only if the identity between the *definiens* and the *definiendum* is self-evidently true. But Frege added that here we really *do not have a definition at all,* since there is no room for us to stipulate the meaning of the symbol in question. The identity presenting the symbol must instead be taken as an axiom.

Unfortunately, Frege does not discuss the logical status of these "definitional" axioms. He very well could have argued that since it is self-evident that the expressions on either side of such identities have the same sense, these axioms are analytic and just as trivial as instances of the law of identity. There would be problems with this argument, however, since there are no noncircular derivations of these axioms from the law of identity: "$a = b$" follows from "$a = a$" only in conjunction with "$a = b$" (or "$b = a$") and a substitutivity rule. Thus, in a formal system these axioms would be needed *in addition* to the usual logical principles and regular definitions. Any proposition whose

[*183*]

proof depended upon them would fail to be analytic according to Frege's *Grundlagen* criterion. In the 1914 essay, however, Frege does not discuss this criterion or propose to modify it.

In the other subcase, the symbol that has been in previous use has no clear meaning. Then it is appropriate to introduce a definition not of the old symbol but of a new substitute for it:

> Assume that A is a simple sign (expression) which has been previously used, whose sense we have tried to analyze logically in that we have formed a composite expression of as presenting this analysis. Since we are not sure whether the analysis succeeds we dare not treat A as replaceable by the composite expression. . . . we must choose a new sign B which is now given a sense for the first time through defining it in terms of the composite expression. Now the question is whether A and B have the same sense. But we can entirely bypass this question by reconstructing the system from its foundation and in so doing no longer use the sign A but rather B.
>
> . . . How is it possible, one can ask, that it can be doubtful whether a simple sign has the same sense as a complex expression. . . . the reason must be that the sense of the simple sign is not clearly grasped, but rather appears only with blurred outlines as through a haze. The result of logical analysis would then be that the sense is clearly presented. This is very useful work; but it does not belong to the construction of the system and must precede it. [Frege 1969:227–228]

Frege's view is thus a forerunner of the Carnapian doctrine of explication. Unlike Carnap, however, Frege deals explicitly only with the sense of an expression. But it certainly would be compatible with his presentation to extend his view to permit reconstructive definitions to fail to preserve reference. There is also some textual support for this. In a letter to Bertrand Russell dated May 21, 1903, Frege remarked that his definition of the real numbers was preferable to Russell's because it obtains both the reals and the rationals in a single step. Since Frege knew that the two definitions are not extensionally equivalent, it seems clear that the preservation of reference was not an important factor in deciding between the two definitions. They were to be judged, instead, on pragmatic grounds.

Frege's critical discussion in the *Grundlagen* makes it quite clear that he thought that prior to his analysis, the concept of number appeared "as through a haze." Thus, from the standpoint of his 1914 essay he would regard his definitions as providing a reconstruction of arithmetic and the concept of number. The question of the analyticity of arithmetic would thus turn upon the nature of this reconstruction. The epistemological status of unreconstructed arithmetic presumably would be dismissed on the grounds that it is incapable of a clear and precisely grounded answer. The problem of analysis need not undermine Frege's logistic program.

Arithmetic

Frege's philosophy of arithmetic consists of three elements: criticisms of opposing views, the analysis of the concept of number, and the reduction of arithmetic to logic. The preceding chapters have covered Frege's criticism of opposing views and their philosophical ramifications, so we will not consider them again here. Let us turn immediately to his analysis of the concept of number.

The Analysis of the Concept of Number

Frege established, through a preliminary analysis, several properties of numbers and discourse about them. The first is that *number is not a property attributable to individual things.* Attributing a number is *not* like attributing a color to a horse, a length to a road, or a weight to a bar of metal. No physical operations are necessary to determine what number a thing to be numbered has. Frege had several arguments in support of this. First, when one says, for example, that there are a thousand leaves on a tree, one ascribes nothing to the individual leaves nor to the foliage as a whole. But when one says that the leaves on a tree are green, one does ascribe properties to the leaves or to the foliage as a whole. Thus, *number* cannot be the same kind of property as *green* (Frege 1884:sec. 22). Second, I can give you a group of things, say a bunch of cards, and ask how heavy they are, and with no further ado

[*185*]

you can give me an answer. But I cannot do the same thing with the question "How many?" For you to answer my question, you need to know what is to be counted. Do I mean how many cards, how many packs, how many suits, or how many pairs? Thus, answers to the question "How many?" are dependent upon something else in a way that answers to the question "How heavy?" or "How long?" or "How fat?" are not (Frege 1884:sec. 22). Third, one and the same situation can give rise to different numbers. I can describe some boots as four boots or two pairs, or two left boots and two right boots. I can describe a stack of cards as two decks or four suits, and so on (Frege 1884:sec. 25). Finally, everything thinkable is numerable. Not only are tables and chairs numerable, but so are ideas, concepts, and theorems. We can count the number of theorems proved by Frege as well as the number of books in his personal library. So number cannot be a physical property, because abstract and mental entities have no physical properties (Frege 1884:secs. 14, 24).

The second point established by Frege's preliminary analysis is that *number words, although they function somewhat like adjectives, are nonetheless different from ordinary adjectives.* He contrasted the word "one" with the word "wise." We can say "Socrates was wise." But "Socrates was one" must be understood as short for something like "Socrates was one of ____," where the blank is filled in by the phrase appropriate to the context in which the statement is made. We can say "Peter and Paul are wise," but we can understand "Peter and Paul are one" only as short for "Peter and Paul are one and the same." Frege continued this linguistic analysis to establish the difference between the word "unit" and the word "one." We say "the number one" and indicate that we have a definite object in mind. But we use the plural with the word "unit," talking about having 5 units of penicillin, for example. This indicates that the word "unit" signifies a general term (Frege 1884:sec. 29).

Linguistic analysis can also establish Frege's third preliminary point: zero and one are numbers. Husserl and other philosophers of Frege's time had argued that they were not. Frege answered them by pointing out that number words are answers to *how many* questions and that certainly "zero" and "one" are answers to *how many* questions. When we ask how many moons the Earth has, the answer is one. When we ask how many kings the United States has, the answer is zero, or none (Frege 1884:sec. 44; 1894:324–328).

Turning from linguistic considerations to more technical matters, let us note that Frege agreed with Leibniz and Mill that the individual natural numbers could be defined in terms of zero and the successor function (Frege 1884:secs. 6, 18). He did not argue in the *Grundlagen* that the class of natural numbers could also be defined as the closure of zero under the successor function, but he already had established this in an earlier paper illustrating applications of logic (Frege 1879a).

As we have noted, he also rejected the geometrical approach to numbers, offered by Newton, according to which numbers are relationships between a given magnitude and a previously selected magnitude taken as a unit, on the grounds that this is too narrow a foundation for arithmetic, since things can be numbered that have no magnitudes of any kind (Frege 1884:sec. 19). Furthermore, we can explain the use of numbers in measurements in terms of counting the number of units of a magnitude to be found in a given magnitude. For example, to say that something is 10 inches long can be interpreted as saying that there are 10 lengths of 1 inch each to be found in the total length of the object. Thus, Frege took number in its use in counting as fundamental, and the other uses as derivative. (In Frege's later geometrical foundation for arithmetic, he reversed himself (1969:298–302).)

Having established the dimensions of the problem, Frege returned again to the fact that the same situation can support different judgments of number equally well. Let us suppose that I have before me a box of cards. With equal right I can say "Here is one deck" or "Here are 52 cards." The difference between the two situations is that I have conceptualized the same physical phenomena differently. But of course Frege could not identify this with my having different ideas. There is something else that varies with the two judgments, the linguistic apparatus used to make them: In the one case I used the predicate "deck" while in the other I used the predicate "card." But number cannot be a property of an expression either qua inscription or qua abstract symbol type. No examination of the expression "deck" will show that I used it to characterize decks of playing cards rather than, say, decks of ships—and surely the usage is relevant to the correctness of my numerical judgment. Frege's *concepts,* on the other hand, fill the bill very well. They are not subjective ideas, nor are they identical with predicates. Rather, predicates refer to them so that a predicate truly characterizes an object just in case that object falls under the concept to which the predicate refers. Varying the linguistic apparatus will vary the predicates and thereby the concepts to which we refer by means of these predicates. But an object falls under a concept whether or not anyone uses language to assert so. Thus, Frege proposed the thesis that *statements of number contain assertions about concepts.*

In ascribing numbers to concepts Frege could guarantee both the objectivity and the linguistic independence of numerical truths. That the Earth has one moon would be true even if no one ever thought or said so; for having the number one is a property not of an idea or a word but of the concept *moon of the Earth* (Frege 1884:sec. 46). And yet the close connection between Frege's concepts and thought and language also explains why we obtain different numerical judgments from the same situation by conceptualizing or describing it differently. It is no accident that Frege's concepts are key ingredients of both his semantics and his philosophy of arithmetic.

This analysis of numerical statements immediately resolves many of the problems that Frege's preliminary investigations had revealed. It explains why such a variety of things can be numbered, for by choosing the appropriate concept we can count both physical and abstract objects as well as collections containing objects of both kinds (Frege 1884:sec. 48). It explains how we can count the number of F's when there are no F's at all, as when we count the number of moons of Venus. For although there are no moons of Venus, there is a concept *moon of Venus* and it is to this that we attribute the number 0 (Frege 1884:sec. 46). Frege's account also resolves the mystery, which had exercised Frege's fellow logicians, of how units are both the same and distinct. In one sense the unit—that is, the thing relative to which we count—remains constant as we count. The unit in this sense is the *concept* being numbered. But in another sense the units counted vary as we count them; these are the objects that fall under the concept (Frege 1884:sec. 54).

Frege's analysis of ascriptions of number would collapse if the distinction between assertions about concepts and assertions about individual things could not be maintained. Thus, it is essential to his account that this concept-object distinction and the attendant distinctions between levels of concepts be an absolute one rather than one that is simply relative to a given analysis of a statement. Frege quite rightly resisted attempts to relativize these distinctions, even though it plunged him into the famous difficulty of the concept *horse* (Frege 1892b; 1884:secs. 51–54).

Having clarified the sense in which ascriptions of number are assertions about concepts, Frege turned to features that might make concepts unsuitable for counting. One of these is that the number of things falling under certain concepts seems to change over time. For example, in 1879, Frege had published *one* major work in logic, while in 1893 his major works in logic numbered *three*. Frege resolved this by arguing that neither the number nor the concept has changed. (Since neither of these exists in space or time, neither can undergo change.) Rather, we have used two different concepts in arriving at our judgments. In the first case we used the concept *major logical work published by Frege on or before the year 1879,* while in the second case we use the concept *major logical work published by Frege on or before the year 1893.* This procedure takes verbs as tenseless and supplies temporal determinations through the addition of an explicit date. Thus, when exhibiting the logical structure of, for example, the statement "Frege discovered quantification theory," we should write "Frege discovers quantification theory on January 10, 1879" (Frege 1884:sec. 46). This move, although a subject of some controversy, has been adopted by many later logicians.

Vagueness can also hinder counting. Lacking sharp criteria for determining whether someone is bald, we may not be able to count the number of bald men in a room. For if we encounter a borderline case, we will be unable to decide how to count him. In actual practice, of course, we may make an arbitrary determination or we may say something like: "We found five bald men and two about which we remained undecided." Frege regarded vague predicates as he regarded proper names without reference: Both must be banned from a logically perfect language. A vague predicate does not refer to a concept on Frege's view. Concepts must satisfy the law of excluded middle: For every object, either x falls under a concept F or it does not (Frege 1903a:sec. 56).

Frege also argued that vague predicates cannot be admitted into logic because they give rise to the sorties fallacy (Frege 1879b:sec. 27; 1896:55). Suppose that a collection of beans is a clear case of a heap. Then if I remove one bean, the collection will still be a heap. In general, if one takes away one bean from a heap of beans, one still has a heap of beans. Suppose that I order the collection of beans in a series, taking the full collection as the first member, and the collection minus one bean as the second, the collection minus two beans as the third, and so forth. Then through induction (or, if one prefers, a finite number of applications of universal instantiation and *modus ponens*), one can show that the last member of this collection is also a heap. But clearly a collection of only one bean is not a heap of beans. One might object that this false conclusion has been drawn from false premises. But which premise do we reject? By assumption, there is no question about the truth of the premise that the original collection is a heap. So, the only choice then is to reject the premise that if a collection of beans is a heap, then when we remove one bean we still have a heap. But if that is false, then its negation is true. But that means that in the series of the collections, there is one collection of beans that is a heap and a one-bean-smaller collection that is not a heap. But this, also, is an implausible conclusion. Frege did not propose a solution, other than counting vague predicates as defects of ordinary language and banning them from his logical language. I shall not pursue the matter further, although I think it is fascinating problem.

A final difficulty connected with counting concerns what we now call *mass terms,* such as "water," "red cloth," and "butter." The trouble here is that if I have a patch of red cloth, a puddle of water, or a mound of butter, and I take away half of the red cloth, butter, or water, I still have something which is red cloth, butter, or water. This prevents me from counting the number of things that fall under the concepts *red cloth, butter,* or *water* without further ado. For how much water do I have? One large puddle or two, three, perhaps even 1 million small puddles, all collected together? In practice, we resolve this problem by counting not water but gallons of water, not butter but pounds of butter, not red cloth but yards of red cloth, and so on. Frege concluded that this shows that when a mass term is taken as the unit of counting, no finite number can be assigned to the concept counted (Frege 1884:sec. 54). If he means to imply, then, that an infinite number can be assigned, this is clearly wrong. First, because exactly the same difficulties would arise in assigning an infinite number to the concept that arise in assigning a finite number to it. The concept by itself just doesn't tell us what to count. Second, given the plausible ways of imposing a counting system on mass terms, it would seem that we would always end up with a finite number rather than an infinite one. For example, suppose that some fantastic turn of events forces me to count the number of things that are water in a particular container of water. Then I will have to decide which parts of the water in the container I am going to take as units to be counted. Clearly, the smallest parts I can count will be molecules of water, since a molecule of water cannot be divided into water in turn. But if I count the molecules of water in the container, I will arrive at a finite number, and similarly if I count any bigger parts of the water in the container.

Frege may have felt pressed to assign numbers to even mass terms, because on his view every function must be defined for all possible arguments and the *number of* function takes any concept as an argument. Thus, as long as mass terms in their predicative use are taken as designating concepts, numbers must be assigned to the concepts they designate. (Numbers can also be thought of as assigned to classes, and in the *Grundgesetze* the *number of* function takes classes as arguments. But these difficulties will still arise with respect to the class of things that satisfy the mass terms.) Frege defined the number of the concept *F* as (roughly) the class of concepts that can be put into one-to-one correlation with the concept *F*. In the case of a concept designated by a mass term (e.g., the concept *butter*), one could argue either that no concept can be put into one-to-one correlation with the concept *butter,* or that only one concept can, the concept *butter* itself. However, the number assigned to the concept *butter* would then be either the empty class or the unit class of the concept *butter*; and in either case it would not be identical with any of the objects that Frege took as the usual numbers, be they finite or infinite.

The difficulties concerning Frege's treatment of the details of the difficult cases do not undermine the valid points in his analysis. When we count the objects on a table, we do not assign a property to the individual objects, nor to the scattered individual composed of them. Furthermore, the number we arrive at is determined by the predicate we use. Recognition of these two points has considerably advanced our understanding of number, whether we agree with Frege that number predicates designate concepts and number assertions are about concepts, or whether we take them to be about classes, as most modern logicians do. Recently, some logicians have proposed treating numerical judgments as syntactically and semantically the same as quantifications, so that the statement "there are five cows in the pasture" is given the same type of analysis as the statement "there are brown cows in the pasture" (Bostock). Frege can agree with at least this part of their views, although he would argue that a quantifier like a number predicate is a higher-order predicate and applies to ordinary predicates.

Frege's second major thesis about numbers is that *numbers are objects*. To understand this, we must remember that for Frege the term "object" is a technical term of his philosophical semantics and stands in opposition to the term "concept." An object is the type of thing that can be referred to by a singular term and can be referred to only by a singular term. A concept, on the other hand, is the type of thing that can be referred to only by a general term. Thus, his thesis that numbers are objects is basically a thesis about the logical form and semantics of number words. He did not mean to say that numbers are concrete individual things, such as tables and chairs. For Frege recognized the possibility of abstract objects as well as concrete ones.

Given his conception of objects, it was natural for Frege to establish his thesis that numbers are objects by introducing syntactical and semantical considerations. He cited two major ones. First, in ordinary language number words function primarily as singular terms. We say "the number five" or "the number of chairs in the room." We don't say "fives" or "number of chairs in the room," except in exceptional circumstances, as when I ask you "How many fives are there?" when I want to know how many groups of five there are in a collection. Furthermore, adjectival uses of number words can be replaced by uses of number words as singular terms. For example, the sentence "five rocks are on the table" can be paraphrased as "the number of rocks on the table is equal to five" (Frege 1884:sec. 57). Frege's second consideration is that in mathematics, numerals function exclusively as singular terms and "our concern here is to arrive at a concept of number usable for the purposes of science; we would not, therefore, be deterred by the fact that in the language of everyday life number appears also in attributive constructions. That can always be got around" (Frege 1884:sec. 57).

One might think that Frege's thesis that numbers are objects is incompatible with his thesis that judgments of number contain assertions about concepts. This apparent difficulty is easily resolved. Number words occur in these judgments not as predicates but as elements of the predicates. This is best seen by paraphrasing numerical judgments in the form of numerical quantifications. For example, we can paraphrase "five men crossed the street" or "the number of men who crossed the street is five" as "there are five men who crossed the street." Then, on Frege's view, we are predicating the higher-level concept *there are five* of the concept *men who crossed the street,* so that the word "five" is simply an element of the predicate rather than the total predicate (Frege 1884:secs. 56, 57). Thus, it is one thing to ask about the reference of "five" and another to ask about the reference of "there are five."

It is not clear from Frege's writings to what extent the thesis that numbers are objects is dependent upon the identification of numbers with extensions of concepts. On the one hand, the analysis of the logical form of numerical terms and statements is sufficient to establish that if there are numbers then they are objects—as opposed to concepts. Also, Frege's arguments that arithmetic is a body of truths, in conjunction with his semantics, are sufficient to establish that numbers exist. On the other hand, when confronted with the Russell paradox, Frege reacted as follows:

> The prime problem of arithmetic may be taken to be the problem: How do we apprehend logical objects, in particular numbers? What justifies us in recognizing numbers as objects? [Frege 1903a:265]

Furthermore, in notes written during the last years of his life, Frege momentarily toyed with the idea that his view that numbers are objects might be based upon a linguistic mistake (Frege 1969:282–283, 277).

Three questions are relevant to this discussion: (1) Are numbers (if there be any) concepts or objects? (2) Do numbers exist? (3) Are numbers logical objects? In grounding number theory in logic (plus the theory of extensions), Frege thought that he had conclusively answered all three questions affirmatively. If numbers are extensions, then it is self-evident that they are logical objects, and that they exist. The Russell paradox called into question the thesis that numbers are logical objects and eventually led Frege to abandon it. However, for some time he evidently did not appreciate the force of his earlier independent arguments for the two other answers. Quite possibly, he thought that if numbers are not logical objects, then there is no alternative—they are not objects at all and his previous account of their logical form was erroneous. But once he decided to base arithmetic upon geometry he shook off those doubts. Numbers once again existed for him as objects—although this time as geometrical objects (Frege 1969:295–299).

Fregean objects are possible references of singular terms. When we refer to an object through a singular term we use the term to pick out that object and distinguish it from other objects. Thus, if our use of the term to refer to an object is to be justified, identities involving the term must make sense. Frege took identity to be a relation whose field is the whole category of objects. Russell suggested to him a theory of types that required dividing the category of objects into subcategories and permitting each concept to take its arguments only from a specified subcategory. But Frege did not think that identity can be accommodated within this framework; for he thought that it is impossible to restrict the range of significance of identity, and he rejected the idea that there is a special identity for each subcategory of objects. "Identity is a relation given to us in such a specific form that it is inconceivable that various kinds of it should occur" (Frege 1903a:254).

The unrestricted range accorded to identity is part and parcel of Frege's treatment of predicates generally. The only semantical categories he recognized are those determined by his hierarchy of functions and objects. Thus, the domain of "is a man" is the whole category of objects, and the sentences "the number two is a man" must be assigned a truth value even if we arbitrarily stipulate it to be false (Frege 1891:18–20). Similarly, the sentence "the number two equals the moon" must have a truth value because identity takes its arguments from the whole category of objects. Frege said that this requirement is imposed upon us by logic, for the law "for every object x, Fx or not-Fx" requires of every first-level concept F that for every object a, Fa is true or false. One might think that this argument could be circumvented by restricting the range of the quantifiers. For example, instead of quantifying over all objects, we could quantify over all numbers or all people or all physical objects, and so on. Then concepts would have to be defined only for the values of the variables that can occur in their argument places, instead of for all objects. But Frege thought that it is in principle impossible to restrict the range of object variables in this way. The universal quantification "for all numbers n, n is an F" must be understood as shorthand for "for all objects x, if x is a number, then x is an F" (Frege 1903a:sec. 66). Given this understanding of the restricted quantification, of course, it is necessary to explain the concept F for all objects as arguments.

Frege's discussion raises several issues. He seems to be arguing that it is impossible to set up a special-purpose formal system whose variables range over a fixed domain. This is, of course, simply false. We have formal systems for elementary number theory whose variables range over only the natural numbers, and the possibility never arises of encountering a concept in such a theory taking something other than natural numbers as its arguments. On the other hand, Frege may be claiming that while it is possible to operate such a symbolism as a formal calculus, it is impossible to understand the sentences in it without presupposing unrestricted universal quantification. Consider, for example, the sentence "for every x there is a y such that $x + y = 0$." This is true if the variables "x" and "y" are interpreted as ranging over the entire set of integers, but false if they are interpreted as simply ranging over the natural numbers. So Frege could argue that in order to achieve an unambiguous understanding of this sentence, we must understand the variables as having an unrestricted range and insert the predicates "is a natural number" or "is an integer." One might concede Frege's point about equivocacy here; but one can counter that the sentence can be disambiguated by taking its variables to range over all numbers rather than over all objects, and that in general, difficulties with restricted quantification can be removed by allowing the variables to range over sufficiently general categories of objects. The trouble here is that no category seems sufficiently general for all purposes, especially when we consider quantification into intensional contexts. For example, I can think of people, numbers, animals, and sets; so in the sentence "there is something x such that I am thinking of x," we must allow the variable to range over all objects. Benacerraf dismisses this example because its intentionality "casts a referentially opaque shadow over the role that identity plays in it" (Benacerraf:66). Thanks to John Gobel I have a nonintensional example to counter with: "Belongs to some set" is a predicate whose range of significance is unrestricted. So also is the predicate "is the fifth member of some series." Given that anything can belong to a set, and that counting the number of members in a set presupposes our ability to distinguish its members, it follows that applied set theory presupposes the universality of identity.

So far we have established (1) that there are theories whose variables range over objects without restriction and (2) that this is not grounded in an artificial stipulation but rather in our presystematic view of the job of the theory in question. This does not establish Frege's view that unrestricted quantification is primary and restricted quantification derivative in the order of our understanding of quantification. We need not resolve this deep issue in the philosophy of logic here. Nor have we demonstrated that the range of the quantifiers and identity must always be taken as unrestricted. Yet, since anything can be counted, even numbers, we see that Frege was correct in the sense that the theory of counting should have an unrestricted universe and identity should be defined even between, say, a number and a chair.

Frege saw that numerical statements of the form "There are n F's" can be explained in terms of first-order logic with identity. For example, the sentence "there are exactly two fish" is true if and only if there are distinct x and y such that x and y are both fish and such that every z that is a fish is identical with either x or y. This idea gives rise to the following scheme (Frege 1884:sec. 55):

$$(\exists 0x)Fx \equiv -(\exists x)Fx$$
$$(\exists n + 1x)Fx \equiv (\exists x)[(\exists ny)(x \neq y \cdot Fy) \cdot Fx].$$

Taking the statement "the number of F's $= n$" as equivalent to the statement "there are n F's," we can use this scheme to eliminate some occurrences of numerals and expressions of the form "the number of F's." But, as Frege saw, the scheme has serious limitations. It is not strong enough to enable us ever to prove that the number of F's is the same as the number of G's even if we know that there are, say, six F's and also six G's. Nor does the scheme explain sentences of the form "the number of F's $= x$" with "x" a variable. The difficulty with the scheme is that it treats numbers not as singular terms but as elements of complex predicates or, if you prefer, numerical quantifiers. For number theory, however, we need an account that introduces numerals as singular terms. The scheme for explaining statements of the form "there are n F's" should be part of our general account but not all of it.

Given that numbers are objects, identity conditions for them must be given. So Frege suggested that we look at their identity conditions for a clue to the explanation of number. Numbers measure the size of classes (for they are a measure of how many members a class has), and two classes are of the same size if and only if their members can be put into one-to-one correlation. So the number of F's is the same as the number of G's if and only if the F's and G's can be put into one-to-one correlation. This gives us identity conditions for numbers. By defining the individual numbers in terms of the *number of* function, we also can explain (1) identities of the form "$n = m$," where n and m are both numerals and (2) those of the form "$n =$ the number of F's," where n is a numeral. For example, suppose that we define 0 as the number of the concept *not self-identical*. Then from our definition and our identity conditions, we obtain: 0 is identical with the number of F's if and only if the F's can be put into one-to-one correlation with the things that are not self-identical. But nothing is not self-identical, so the F's will be in one-to-one correlation with things that are not self-identical if and only if there are no F's. So, the number of F's is 0 if and only if there are no F's, and our definition is satisfactory. Given the definition of 0, we can then define the number one as the number of the concept *identical to 0*. Since only one thing is identical to 0, it follows that the number of F's equals 1 if and only if the F's can be put in one-to-one correlation with the things which are identical to 0 (i.e., if and only if there is exactly one F). In a similar way, we can go on and define two as the number of the concept $x = 0$ *or* $x = 1$, and in general define the number $n + 1$ as the number of the concept $x = 0$ *or* $x = 1$ *or* . . . *or* $x = n$. We can even define the successor predicate as follows: m is the successor of n if and only if there is a concept F such that m equals the number of F's and there is an object x which is an F and n is equal to the number of F's distinct from x. Using this definition and the theory of one-to-one correlations, one can prove that the successor relation is a one-to-one relation. Indeed, we can also use Frege's technique and define the natural numbers to be the smallest class containing 0 and closed under the successor relation, and in this way obtain a model for the Peano axioms.

Yet this procedure does not succeed in defining the *number of* function. It succeeds only in defining it for contexts of the form "the number of F's $= t$," where "t" is either a numeral or expression of the form "the number of G's."

On the other hand, if we can explain contexts of the form "the number of F's $= y$" with "y" a variable, we can eliminate all occurrences of *number of* expressions, no matter what context in which they occur. For one of the laws of quantification theory with identity is that "Fa" is equivalent to "$(\exists y)(a = y \cdot Fy)$," and this allows us to eliminate all occurrences of expressions of the form "the number of F's" in favor of their occurrences within identities only. It might be suggested that we handle this last case by stipulating that the "the number of F's $= y$" is false when y is not a number. The problem here is that this presupposes exactly the concept of number we are trying to define. Indeed, given the approach we are taking, the most natural way to define the concept of number is as follows: x is a number if and only if $(\exists F)(x = $ the number of F's) (Frege 1884:secs. 56, 72). Furthermore, even when y is a number, we can determine the truth value of "the number of F's $= y$" only when y is known to us as the number of some particular concept. For only then can we apply our identity conditions for numbers. But, of course, in a general theory of counting, there is no reason to suppose that every number will be given to us in this form. (Cf. Parsons.)

These considerations led Frege to reject a definition of number that proceeds to define them contextually through the identity context and their identity conditions (Frege 1884:sec. 56). In the *Grundlagen* he does not seem to be opposed to contextual definitions in principle, but in his *later* works he was emphatic about rejecting them (Frege 1903a: secs. 66–69, p. 255). Although Frege rejected contextual definitions before they were known by that name and before Russell had introduced them to the mathematical and philosophical world, I think that he would reject them even if he were to write today. This is because the unfettered use of contextual definitions requires a metatheoretic investigation establishing that the singular terms which they introduce can be freely substituted for the variables of quantification. Not only is this against the spirit of Fregean definitions in the object language, but it also is against his principle that no definition should depend upon the proof of a mathematical theorem for its correctness (Frege 1903a:sec. 60; 1903b:370).

The Technical Solution

One-to-one correlation is an equivalence relation, and therefore it partitions its field into equivalence classes. Furthermore, if E is any equivalence relation, then an E-equivalence class containing a is identical to an E-equivalence class containing b if and only if a bears E to b. As a particular case, the number of F's will be identical to the number of G's if and only if the F's and G's can be put into one-to-one correlation, provided that we identify the number of F's with the class of things in one-to-one correlation with the concept F. So Frege took as his definition: The number that belongs to the concept F is the extension of the concept *in one-to-one correlation with the concept F*. In the *Grundgesetze* this was amended so that the number of a class α is the class of all classes in one-to-one correlation with α. I have argued elsewhere that careful exegesis will show that these two definitions are essentially the same (Resnik 1965). To disentangle Frege's formal definition of number from his theory of concepts and objects, and to use modern terminology and notation, I will make use of the latter definition in my discussion.

It is only natural to ask whether Frege's definition has accomplished what he expected from it. In fact, the definition satisfies the desiderata which Frege set forth, together with others proposed by subsequent students of the concept of number. Frege required, you will recall, that his definition allow us to derive the identity conditions for numbers, namely, that the number of the class α is identical to the number of the class β if and only if α and β can be put into one-to-one correlation. This is an immediate corollary of his taking numbers as equivalence classes with respect to the relation of being in one-to-one correlation. Frege sketched the easy proof of this in the *Grundlagen* (Frege 1884:sec. 73).

Another principle which arose in Frege's analysis is that the number of a class α equals n if and only if $(\exists nx)(x \in \alpha)$. Owing to the use of the numerical quantifier, the letter "n" must be interpreted as a schematic letter for numerals rather than as a variable ranging over numbers. Thus, the theorem really is a metatheorem. Once the definitions of the individual numbers are given—either in terms of successor or, as Frege indicates in the *Grundlagen*, in terms of the class of all earlier numbers, the theorem can be proved in a fairly straightforward manner by metalinguistic induction. Frege did not prove this theorem and proved only two of its instances—the cases of 0 and 1 (Frege 1893).

The problem of establishing the relationship between numbers and counting also can be approached by defining "α has n members" by "α is in one-to-one correlation with the numbers less than n." Then "n" can be taken as a variable, and the theorems mentioned above proved as universal quantification over all numbers. An adequate account of counting certainly should include these laws. For to count a finite class is to number each of its members, and that establishes a one-to-one correlation between the class and the class of numbers up to a given number. Ordinarily, of course, we count starting with the number 1 rather than with the number 0, so that the last number used in the counting process will be the number of the class counted. But this is not an essential feature of counting. We could count just as well starting with the number 0 and then add 1 to the last number used in counting the class in order to obtain the number of the class. Frege does not state these features of counting explicitly, although they are implicit in his approach.

Any adequate analysis of number must provide also for the theorems of number theory. When Frege wrote the *Grundlagen* in 1884, number theory had not been axiomatized. By 1893, when he wrote the first volume of the *Grundgesetze,* Dedekind's analysis had been published, and Frege explicitly refers to it. Dedekind's work contains the first explicit formulations of the so-called Peano axioms. Although Dedekind claimed that these axioms characterized the natural number sequence, he did not claim explicitly, at least, that they are sufficient for every theorem of number theory (Dedekind). Frege established each of these axioms in the *Grundgesetze,* but he did not gather them together under one heading, nor did he claim that they are sufficient for number theory. It just never seemed to occur to Frege to argue that number theory reduces to logic *because* a set of axioms sufficient for number theory can be modeled or translated into logic. Instead, his program, like that of Russell and Whitehead, seems to have called for a detailed proof in logic of every important theorem of arithmetic. This was necessary in *any case* because nobody had given such detailed proofs from any axioms. But as the more abstract concept of a mathematical reduction had not yet been introduced, Frege probably never thought of these problems in the terms that contemporary logicians would use.

More recently, Paul Benacerraf has proposed another condition which any adequate definition of number should satisfy. He notes that when a child learns to count he learns two things: (1) how to assign numbers to collections of objects, and (2) how to generate the number series without counting anything while he is doing it. In other words, the child learns how to answer questions such as "How many fingers am I holding up?" but also learns to respond to commands such as "Count to 110."

Benacerraf calls the generative type of counting *intransitive counting,* and points out that in order for an analysis of number to explain the possibility of intransitive counting by human beings, the notation for the number series should be generated effectively (Benacerraf:50). Either of the two schemes suggested by Frege's writings will meet this condition. For if the individual numbers are taken as 0, $S(0)$, $SS(0), \ldots$, and 0 and the successor function are given Frege's explicit definitions, we will have an effective set-theoretic notation for generating names of the numbers. On the other hand, if the numbers are given by

$$0 = \text{number of } \hat{x}(x \neq x)$$
$$S(n) = \text{number of } \hat{x}(x \leq n)$$

we can again obtain an effective notation by means of Frege's definitions.

Another important piece in Frege's technical reduction is the definition of the ancestral of a relation. A person x is an ancestor of a person y if and only if x is a parent of a parent of a parent of . . . of a parent of y. In other words, just in case there are z_1 through z_n, such that x is a parent of z_1, z_1 is a parent of $z_2, \ldots$, and z_n is a parent of y. This explanation of the ancestor relation in terms of the parent-of relation cannot be taken as a definition because it makes use of the dots. Frege showed how to get a proper definition using second-order logic (Frege 1879b:secs. 23–26; 1884:sec. 81). To understand definition, let us call a property F *hereditary* just in case for every x and y if x has F and is a parent of y, then y has F. In other words, hereditary properties are those passed from parent to child. If x is an ancestor of y and x has a hereditary property, then y must also have it. Thus, a necessary condition for x to be an ancestor of y is that y have every hereditary property that x has. If we count a person as one of his own ancestors, then this condition is sufficient, too. For suppose that y does have every hereditary property which x has. Since *having x as an ancestor* is hereditary and x has this property, y must also have it. But that means that x is an ancestor of y. Thus, we may define

$$x \text{ ancestor } y \equiv (F)[(Fx \cdot (z)(w)(Fz \cdot z \text{ parent } w \supset Fw)) \supset Fy],$$

which gives us a definition in second-order logic of "ancestor" in terms of "parent."

[*202*]

Frege's definition was not quite this simple, however, because he did not count a person as one of his own ancestors. Instead, he defined what is now known as the *proper ancestor* relation by the condition: y has every hereditary property which x's children have. In symbols, this runs as follows:

$$x \text{ proper ancestor } y \equiv (F)[((z)(w)((Fz \cdot z \text{ parent } w) \supset Fw)$$
$$\cdot (u)(x \text{ parent } u \supset Fu)) \supset Fy].$$

Frege saw that this technique can be generalized to any relation R so that the *ancestral* of R can be defined in second-order logic according to the same format. In particular, we define *R-hereditary* by

$$F \text{ is } R\text{-hereditary} \equiv (x)(y)((Fx \cdot xRy) \supset Fy)$$

and then the R-ancestral by

$$x \text{ } R\text{-ancestral } y \equiv (F)[(Fx \cdot F \text{ is } R\text{-hereditary}) \supset Fy].$$

This definition can be applied to define the class of natural numbers in terms of 0 and the successor relation, and to derive the principle of mathematical induction. Let "xSy" abbreviate "y is the successor of x" or "x has y as its successor." A natural number is 0 or the successor of 0 or the successor of the successor of 0, and so on. So x is a natural number just in case 0 is a S-ancestor of x.

The principle of mathematical induction states that if 0 has a property F and a natural number has F only if its successor does, then every natural number has F. To derive this from our definitions, let us assume that the antecedent condition is true and that x is any natural number. That is, we have

(1) $F0$
(2) F is S-hereditary
(3) 0 is a S-ancestor of x.

But then from the definition of S-ancestor, we have

(4) $(F)(F0 \cdot F \text{ is } S\text{-hereditary} \supset Fx)$.

So "*Fx*" follows by universal instantiation with respect to "*F*" and truth-functional logic. Frege's definition can also be formulated in terms of classes, which avoids the use of second-order logic. (We define *x* is an ancestor of *y* if and only if *y* belongs to every class that is closed under the parent-of-relation, where a class α is closed under the parent of relation if and only if for every *x* and *y* if $x \in \alpha$ and *y* is a parent of *x* then $y \in \alpha$.)

A final detail necessary to complete the exposition of Frege's technical analysis is the definition of one-to-one correlation, since one might argue that this is not a logical notion but a mathematical one. Frege answered this objection by pointing out that a class α and a class β are in one-to-one correlation if and only if there is a relation *R* that matches one-to-one the objects in α with those in β. But this means first that *R* is a matching relation and second that for every member *x* of α there is a member *y* of β such that *x* bears *R* to *y*, and conversely for every member *z* of β there is a member *w* of α such that *w* bears *R* to *z*. This last condition is formulated in logical terms, so we need only show that the property of being a matching relation is a logically definable one. This is easy to do, because *R* is a matching relation if and only if (1) $(x)(y)(z)(Rxy \cdot Rxz \supset y = z)$ and (2) $(x)(y)(z)(Ryx \cdot Rzx \supset y = z)$ (Frege 1884:secs. 70–72).

Set Theory

The success of Frege's analysis depends upon his view of classes. For only if classes are logical objects and class theory is a branch of logic can it be claimed plausibly that he has shown that number theory or other parts of arithmetic is reducible to logic. Frege seems to have thought that this hardly needs arguing, and I detect three reasons for his viewing the matter in this light. First, according to the tradition to which Frege belonged, the theory of classes belongs to logic. Frege took it for granted that class theory belongs to logic, just as we take it for granted that the theory of the conditional does. If anything, he thought that previous logicians had put classes in the wrong place by attempting to base predicate and propositional logic on them (Frege 1893:VII; 1912:251; 1969:16–20, 36–39, 51–59). Second, he thought that the only legitimate notion of class is one that construes classes as extensions of concepts, thereby placing them firmly in the domain of logic (Frege 1895). Third, the generality argument used to make it plausible that arithmetic rests upon logic also transfers to class theory (Frege 1885:94–96).

As far as the argument from tradition is concerned, there is little to say about it except that Frege's analysis has given rise to a *new* tradition according to which it is questionable that classes or sets belong to logic. For Frege showed us how to develop at least first-order logic without committing ourselves to classes. Ironically, Frege himself was aware that strong commitments to classes arise not in founding elementary logic but rather in the development of arithmetic (Frege 1912:251).

The generality argument is a double-edged sword. Without doubt, anything can belong to a class, so that class theory is as general as logic in terms of its potential applications. On the other hand, the class theory necessary to produce number theory requires much stronger existential assumptions than does first-order logic, and the contrast is even more striking if the set-theoretic foundations for real number theory and analysis are also included. For first-order logic holds in any nonempty universe, while the class theories in question require infinite universes for their truth. Thus, first-order logic holds title to much stronger claims to the sort of general framework theory that logic is usually taken to be.

The status of second-order logic is somewhat ambiguous with respect to this discussion because Frege interpreted second- and higher-order variables as ranging over functions, concepts, and relations, none of which he identified with sets or classes. Today, however, these Fregean notions have been abandoned, and second- and higher-order logics have been interpreted as carrying commitments to sets. Thus, it has been argued that second- and higher-order logics have no clear title to a place in the domain of true logics. (See Boolos 1975 for a full discussion of this issue.)

Frege's conception of classes as extensions of concepts is of particular interest, since it not only led to the Russell contradiction but is also opposed to the iterative conception of sets which dominates the mathematical scene today. The iterative conception views set theory as dealing with a mathematical structure whose members—the sets—form an iterative hierarchy generated by applying the *set of* operation. At the lowest level is the null set, at the next level is the null set together with all sets formed from it (i.e., its unit set), and the process of forming sets thus from previously formed sets is iterated indefinitely. (For a fuller exposition, see Boolos 1971.) Of course, sets are not formed literally in generating the hierarchy. The whole structure is best viewed as already on hand, with set formation being a heuristic notion that enables us to learn how the sets in the hierarchy are related to each other.

This conception of set traces back to two notions that were prevalent during Frege's time. One is the mental collection notion, which may be found in the works of Cantor and Husserl. This views sets as created somehow by a mental gathering together, and is subject to most of Frege's anti-ideational arguments. The other is the aggregate view of sets, which may be found in Schroeder's work. This treats sets as compound individuals whose "members" are their parts. It is formalized best through the calculus of individuals, and, as Frege showed, does not furnish us with a notion of set adequate to the needs of logic or mathematics. In particular, he noted that this approach is unable to furnish a null set (since every individual is part of itself) nor able to recognize the intransitivity of set membership (Frege 1895).

The faults with Schroeder's and Cantor's conceptions are tied to their particular expositions and do not vitiate the modern iterative conception. Nonetheless, this conception does not suit Frege's purposes; for it takes the set-theoretic hierarchy as a mathematical structure on a par with other mathematical structures, such as the natural number sequence and Euclidean space. It is not clear at all how such a structure can be known via logical means, nor do the axioms of the iterative theory, which postulate particular sets and modes of forming sets, seem to be candidates for principles of logic. The topic neutrality, which is the prime characteristic of logic, is missing in the iterative conception.

Frege needed sets or their surrogates to provide the "logical objects" necessary to found arithmetic. He was aware he could meet all the other needs of logic with just the two truth-values and concepts and relations:

> But numbers are objects, and in logic we have only two objects
> in the first place: the two truth-values. Our first aim, then, was
> to obtain objects out of concepts, namely, extents of concepts or

classes. . . . The difficulties which are bound up with the use of classes vanish if we only deal with objects, concepts, and relations, and this is possible in the fundamental part of logic. The class, namely, is something derived, whereas in the concept—as I understand the word—we have something primitive. [Frege 1912:251]

Thus, Frege's quest for logical objects led him to tie classes to concepts. There were from his point of view several advantages to this. Concepts are not mental or linguistic-dependent. Thus, objects associated with them would have the human independent existence needed for numbers. Furthermore, there are empty concepts (i.e., concepts which apply to nothing), and most concepts do not apply to the parts of the objects to which they apply. Thus, the difficulties with Schroeder's domain calculus could be avoided.

The problem of introducing these logical objects in a logically impeccable manner still remained. There Frege returned to an idea developed in the *Grundlagen:* We can introduce logical objects if we can fix identity conditions for them. It is only natural to take the relationship between concepts and logical objects to be a one-to-one relationship, but there is a problem: Identity (which is used in characterizing one-to-one relations) only holds between objects. But for concepts there is an analogous relationship: $(x)(\phi x = \psi x)$ (Frege 1969:132). Thus, if we let "$\hat{x}Fx$" represent the logical object associated with the concept F—its extension—the one-to-one relationship may be formulated by

$$(1) \quad (\hat{x}Fx = \hat{y}Gy) \equiv (x)(Fx \equiv Gx),$$

which is Frege's famous axiom V.

This buys us quite a bit. The identity between two extensions or classes, at least when they are presented as such, can be settled by examining the concepts with which they are associated. Furthermore, even class membership can be reduced to a question about concepts since

$$(2) \quad a \in \hat{x}\,Fx \equiv (\exists G)[(x)(Gx \equiv Fx) \cdot Ga].$$

The result is that most questions about classes that arise in the context of class theory can be reduced to questions about concepts, provided that we accept the two principles (1) and (2). In Frege's system, the latter turns out to be a definitional consequence of the former. Thus, one axiom—principle (1)—can "reduce" class theory to concept theory and thereby to logic.

Unfortunately, (1) is not strong enough—forgetting for the moment that it yields the Russell paradox when combined with Frege's other axioms—for (1) does not determine whether $\hat{x}Fx$ is identical to an object y when y is not given as a class abstract. This is the same difficulty that led Frege to reject a similar principle as a definition of number. But Frege is not defining class abstracts here, he is introducing them as primitives. Thus, he is free to leave these questions formally unsettled as long as the semantics for class abstraction do not preclude a full interpretation of his symbolism. In section 10 of the *Grundgesetze,* Frege also concerned himself with another indeterminacy in his treatment of classes: if h is a one-to-one function—at least when restricted to classes—then, by (1),

$$h(\hat{x}Fx) = h(\hat{y}Gy) \cdot \equiv (x)(Fx \equiv Gx),$$

and if h maps classes onto nonclasses, then classes have the same identity conditions as some other type of objects. But classes have been introduced only through their identity conditions, and thus we would be unable to distinguish them from other entities (if any) also having these identity conditions.

Frege thought that these difficulties could be overcome by fixing values for every function introduced in the system, which takes classes as arguments. His reasoning is not explicit but seems to be this: (a) since the identity relation is a function of the system, the determination of its values (which are truth-values) will settle questions concerning whether classes are identical to objects of seemingly other types, and (b) since the determination will also fix a reference for every complex name of the form "$F(\hat{x}Gx)$," it will thereby guarantee that class abstracts are referential. (Cf. Frege 1893:sec. 29–31.) At the point where this discussion occurs, Frege had only introduced functions into the system which have truth-values as values. He showed that their application to classes can be reduced to the application of $\xi = \zeta$ to classes. This he accomplished by construing each of these truth-value-yielding functions as a composite function consisting of a truth function applied to the horizontal function. For example, the negation function $\neg \xi$, which yields the false for the true as an argument and the true for every other argument, may be construed as $\neg(\text{———}\xi)$, where $\text{———}\xi$ yields the true for the true as an argument and the false otherwise. Thus, the direct application of a truth-function to a non-truth-value may be replaced by its application to a value of the horizontal function. But $\text{———}\xi$ is in turn equivalent to $\xi = (\xi = \xi)$. Thus, everything can be reduced to determining the truth of identities involving classes and possibly other objects.

The only other objects that Frege had introduced into his ontology were the truth-values. So he concluded that he only had to decide whether truth-values were classes. Principles (1) and (2) do not decide this, of course, so the matter must be settled by stipulation. Frege then proceeded to show that it is consistent to stipulate that the truth-values are two distinct classes. His argument can be put into more contemporary terms as follows. Suppose that M_1 is a model of (1) whose domain is restricted to classes and which has at least two distinct members $\hat{x}F_0 x$ and $\hat{y}G_0 y$. Extend the domain of M_1 to a new domain by adjoining the two truth-values and define a function f by $f(\hat{x}F_0 x) = T$, $f(T) = \hat{x}F_0 x$, $f(\hat{y}G_0 y) = F$, $f(F) = yG_0 y$, and $f(x) = x$ for all other x. Then f is a one-to-one function that permutes the extended domain and in which

$$f(\hat{x}Fx) = f(\hat{y}Gy) \equiv (x)(Fx \equiv Gx)$$

is satisfied. So if we interpret "$\hat{x}Fx$" by assigning it $f(\hat{x}Fx)$ we can obtain a model M_2 of (1) in which the two truth-values are "classes."

Frege then stipulated that the true and the false are two distinct classes, so that all objects in his ontology would be classes. In introducing later functions he made sure that their values would also be classes.

Unfortunately, this procedure does not resolve all the difficulties it was designed to handle. Since the identity condition for classes can be applied only to classes that are presented via abstracts, identities of the form "$\hat{x}Fx = y$," where "y" is a variable, are not given a truth-value for every assignment to "y." Furthermore, a class abstract of the form

$$\hat{y}(F)(y = \hat{x}Fx)$$

will be referential only if "$\hat{x}Fx$" designates a class for each assignment to "F." Thus, Frege's argument (Frege 1893:sec 31) that "$\hat{x}Fx$" is referential provided that "$F\xi$" is a referential function *name* is *too weak* to show that abstracts of the foregoing form are referential. (This was first noted in Bartlett.) It works only for "predicative" abstracts (i.e., those which do not involve quantification over all concepts).

Frege could have avoided this problem if he had simply introduced the abstraction operator as a second-level function which yields a class when applied to any first-level function. Thus, it is natural to look for a deeper reason behind his circuitous method. Since Frege attempts to secure references for class abstracts by fixing a reference for every complex name containing them, rather than by the direct assignment of classes to abstracts, the context principle appears still to have had some hold over him (Snapper). But before we conclude that the context principle is at work here, let us note that *even if Frege had used the direct method, he still would have been faced with the same indeterminacy problems*. For even if we introduce classes directly and do not tie them to class abstracts, as is done in modern set theories, it will still be left undetermined whether concrete individuals or truth-values are classes. These questions can be bypassed by restricting variables for classes to classes only, but those who wish to follow Frege or Quine in having unrestricted individual variables must expect to use stipulations to settle them. (Cf. Quine 1963:29–33.) Thus, Frege's procedure in section 10 of the *Grundgesetze* is not attributable entirely to his desire to provide a contextual interpretation of class abstraction.

Furthermore, in section 9 Frege seems quite comfortable with a direct explanation:

> I write "$\grave{\epsilon}(\epsilon^2 - \epsilon) = \grave{\alpha}(\alpha \cdot (\alpha - 1))$," in which by "$\grave{\epsilon}(\epsilon^2 - \epsilon)$" I understand the course-of-values of the function

[*210*]

$\xi^2 - \xi$, and by "$\grave{\alpha}(\alpha \cdot (\alpha - 1))$" the course-of-values of the function $\xi \cdot (\xi - 1)$. Similarly, $\grave{\epsilon}(\epsilon^2 = 4)$ is the course-of-values of the function $\xi^2 = 4$, or, as we can also say, the extension of the concept *square root of* 4. [Frege 1893:sec. 9]

Thus, it seems just as likely that Frege was *not* attempting a contextual interpretation of class abstraction and was rather attempting to guarantee that the functions he had introduced, including the one that yields the course-of-values of a function, would be everywhere defined. He mistakenly assumed in his proof that a function f is everywhere defined provided that "$f(a)$" has a reference for every referential *name a*. But if there are unnamed entities in the domain of f, f could satisfy Frege's condition and fail to be defined for one or more unnamed arguments. He applied this assumption to all his other primitive function names as well, and there is no hint of a contextual introduction of these.

Criticisms of Frege's Theory

A number of logical and philosophical objections can be leveled against Frege's views. I shall discuss Frege's philosophy of arithmetic by starting with specific and technical points and then proceed to general philosophical issues.

The Contradiction in Frege's System

Shortly before Frege published the second volume of the *Grundgesetze,* he received a letter from Bertrand Russell explaining the famous paradox of the class of all classes that are not members of themselves. Frege immediately saw that this gave rise to a contradiction in his formal system and was a disaster for his foundations for arithmetic and set-theoretic reductions generally. To understand how pervasive the effects of the paradox are on Frege's system, we must consider the paradox in detail. In doing this, however, I shall make use of modern notation and terminology. Also, unlike Frege, I will distinguish identity from the biconditional, since Frege's fusing of them is not an essential feature of his set theory. With these qualifications, his underlying logic can be taken as the second-order predicate calculus with identity. Frege's individual variables range over Fregean objects and are disjoint from predicate variables that range over Fregean concepts. This feature avoids the predicate version of the Russell paradox because the sentences "$F(F)$" and "$-F(F)$" are malformed. However, although Frege's logic is a type theory, he broke type distinctions through his introduction of sets. This was accomplished by taking class abstraction as a primitive operator and allowing it to apply to any open sentence in the notation of the system.

On Frege's approach, the membership relation is not taken as primitive but rather is defined by

$$x \in y \equiv (\exists F)(y = \hat{y}Fz \cdot Fx).$$

Classes are then governed through Frege's famous axiom V:

$$\hat{x}Fx = \hat{y}Gy \equiv (z)(Fx \equiv Gz).$$

Since, $\hat{x}Fx = \hat{x}Fx$, $Fa \supset (\hat{x}Fx = \hat{x}Fx \cdot Fa)$; so, by second-order logic,

(1) $(a)(Fa \supset a \in \hat{x}Fx).$

Conversely, by definition, $a \in \hat{x}Fx \supset (\exists G)(\hat{x}Fx = \hat{y}Gy \cdot Ga)$. Thus, by axiom V and second-order logic,

(2) $a \in \hat{x}Fx \supset Fa.$

Thus, by (1) and (2) we get the usual abstraction principle,

(3) $a \in \hat{x}Fx \equiv Fa.$

Then by substitution of "$\hat{x}(x \notin x)$" for "a" and "$x \notin x$" for "F" in (3), we get the Russell contradiction,

$$\hat{x}(x \notin x) \in \hat{x}(x \notin x) \equiv \hat{x}(x \notin x) \notin \hat{x}(x \notin x).$$

[*212*]

Frege's derivation of the paradox dispensed with the use of membership and (3). Let us define the concept R and the class r by

$$Rx \equiv (\exists F)(x = \hat{y}Fy \cdot -Fx)$$
$$r = \hat{x}Rx.$$

Then by the definitions and second-order logic,

$$(4) \quad -Rr \supset (F)(r = \hat{x}Fx \supset Fr).$$

But again by second-order logic,

$$(5) \quad (F)(r = \hat{x}Fx \supset Fr) \supset (r = r \supset Rr).$$

The condition "$r = r$" can be dropped, so by (4) and (5),

$$(6) \quad -Rr \supset Rr:$$

thus,

$$(7) \quad Rr.$$

By definition, (7) yields

$$(8) \quad (\exists F)(r = \hat{x}Fx \cdot -Fr).$$

Let F be any concept satisfying (8); that is,

$$(9) \quad r = \hat{x}Fx \cdot -Fr.$$

Then, by axiom V,

$$(10) \quad (x)(Rx \equiv Fx).$$

So

$$(11) \quad -Rr,$$

contradicting (7).

Frege's axiom V implies two conditionals:

(Va) $(z)(Fz \equiv Gz) \supset \hat{x}Fx = \hat{y}Gy$
(Vb) $\hat{x}Fx = \hat{y}Gy \supset (z)(Fz \equiv Gz)$.

The first is unobjectionable, since it only postulates the extensionality of classes. But the second is used in deriving (11). Notice that (7), the other half of the paradox, is a theorem of second-order logic. So the avoidance of the paradox must come through a modification of (Vb). If we view the abstraction operation as Frege did, as a function that maps concepts into objects, then (Vb) states that this function is one-to-one. But there can be no one-to-one function of this type, since Cantor's theorem implies that there are more sets of objects than there are objects. So the paradox was to be expected. Indeed, in the appendix to the *Grundgesetze*, Frege generalized the paradox to establish that there is no such one-to-one function. This yields the result

$$(\exists F)(\exists G)[\hat{x}Fx = \hat{y}Gy \cdot -(z)(Fz \equiv Gz)].$$

Thus, if two classes coincide, some objects might fall under the concept of one and not under the other. Frege argued that these objects are the classes themselves and thus replaced (Vb) by

$$(V'b) \; \hat{x}Fx = \hat{y}Gy \supset (z)(z \neq \hat{x}Fx \supset (Fz \equiv Gz)).$$

Then in the derivation of (11), step (10) is replaced by

$$(10') \; (x)(x \neq r \supset (Rx \equiv Fx)).$$

This blocks the passage to (11), since we cannot prove that ''$r \neq r$.''

Frege's repair was made in haste and under the pressure of a publication deadline. He had time to check a few of his original proofs to see that they could be modified to accommodate (V'b), but he did not have time for a thorough investigation of its consequences. In his surviving manuscripts there are no further developments of his set theory, al-

[*214*]

though I have been told by J. M. Bartlett that a manuscript of a third volume of the *Grundgesetze* was destroyed during World War II. In notes written in his last years, Frege expressed considerable distrust of set theory and in the end sought a geometrical foundation for arithmetic (Frege 1969:295–302).

Subsequent writers have shown that Frege's repair does not work because a new contradiction can be derived in the presence of the additional assumption that there are at least two objects. Lesniewski first saw this in 1938 (Sobocinski). But his argument is not easily extracted from his peculiar approach to set theory. Later Quine and Geach presented contradictions derived within the farmework of standard set theory (Quine 1955; Geach). However, the motivation behind their proofs is hard to detect, and Geach uses a stronger axiom than Frege's. (Geach's axiom is implicit in Frege's remarks, however.) For this reason I will present my own demonstration of the failure of Frege's repair.

Let us first note that the Russell paradox can be generalized. This is achieved by defining the concept R_f and the class r_f by

$$R_f x = (\exists F)(x = f(\hat{y}Fy) \cdot -Fx)$$
$$r_f = \hat{x}R_f x.$$

Here f represents an arbitrary function defined on classes, such as the unit class function or the complement function. Using steps (4) through (7) with the appropriate substitutions, we can easily prove

(7') $R_f f(r_f)$.

In proving "$-R_f f(r_f)$" we are blocked from passing beyond

(9') $f(r_f) = f(\hat{x}Fx) \cdot -Ff(r_f)$

unless we have

(12) $(x)(y)(f(x) = f(y) \supset x = y)$.

But this just requires f to be one-to-one, and both the unit class and complement functions satisfy this. So for every one-to-one function we have a Russell paradox (Bartlett).

In the repaired system we are also blocked from passing beyond

$$(10') \ (x)(x \neq r_f \supset (R_f x \equiv Fx))$$

even when f is one-to-one, for we need

$$(13) \ f(r_f) \neq r_f.$$

This follows from

$$(14) \ (x)(f(x) \neq x).$$

Together, (12) and (14) require f to be a one-to-one function without fixed points. Such a function always can be defined on any domain of two or more individuals, given the axiom of choice. So Frege's repaired system has no *standard* models in domains with more than one individual.

The result we have gotten so far is weaker than Geach's, Lesniewski's and Quine's; for the system might have a nonstandard model, that is, one on which not every function of the individuals is in the range of the function variables. (See Henkin for the definition of standard model.) So let us see if we can construct a function in Frege's repaired system for which (12) and (14) hold.

From the definition of "∈," we immediately get

$$(15) \ Fa \supset a \in \hat{x}Fx.$$

So, defining "V" as "$\hat{x}(x = y)$" and "ix" as "$\hat{y}(x = y)$," we have

$$(16) \ (x)(x \in V)$$
$$(17) \ (x)(x \in ix).$$

From (Vb) and the definition of "∈," we obtain

(18) $a \neq \hat{x}Fx \supset (a \in Fx \supset Fa)$.

This yields

(19) $(y)(y \in ix \supset (x = y \lor ix = y))$
(20) $(x)(x \in \Lambda \supset x = \Lambda)$,

where "Λ" is defined by "$\hat{x}(x \neq x)$." Notice that by (19) a "unit class" has at most two members, and by (20) the "null class" has at most one.

Assume that $ix = iy$. Then, by (17), $y \in ix$. So if $y \neq iy$, then $y \neq ix$ and, by (19), $x = y$. Discharging our assumption, we get

(21) $ix = iy \supset (y \neq iy \supset x = y)$.

Again assume that $ix = iy$. Then if $x \neq y$ and $iy = y$, we have $iy \neq x$. But by (17) and the assumption, $x \in iy$. So, by (19), $x = y$. Discharging assumptions yields

(22) $ix = iy \supset (y = iy \supset (x \neq y \supset x = y))$.

So by truth-functional logic, we have

(23) $ix = iy \supset x = y$,

which shows that the unit class function is one-to-one even in Frege's repaired system.

We now add to Frege's repaired system the assumption

(24) $\Lambda \neq V$.

(This is equivalent within the system to "$(\exists x)(\exists y)(x \neq y)$.") Next we define the function f as the union of

$f_1(V) = \Lambda$
$f_2(\Lambda) = V$

$$f_3(x) = ix \text{ if } x \neq ix,\ x \neq \Lambda,\ x \neq V$$
$$f_4(x) = \{x, \Lambda, V\} \qquad \text{if } x = ix,\ x \neq \Lambda,\ x \neq V.$$

If $x = ix$, then x has at most two members while $f_4(x)$ has at least three, by (24). So

(25) $(x)(f(x) \neq x)$.

We have shown already that ix is one-to-one, and a straightforward proof shows that $\{x, \Lambda, V\}$ is one-to-one when restricted to x's such that $x \neq \Lambda$, $x \neq V$, and $x = ix$. Thus, to show that f is one-to-one we need only prove that the ranges of f_1 through f_4 are disjoint. This amounts to establishing that

(a) $V \neq \Lambda$ (this is (24))
(b) $V \neq ix$ if $x \neq \Lambda$, $x \neq V$, and $x \neq ix$.

If $x \neq ix$, then by (19) only x belongs to ix, while Λ, $x \in V$, and (b) follows.

(c) $V \neq \{x, \Lambda, V\}$ if $x \neq \Lambda$, $x \neq V$, and $x = ix$.

We shall prove that iV is distinct from x, V, Λ, and $\{x, V, \Lambda\}$ under the conditions of (c). Then $V \neq \{x, \Lambda, V\}$ since $iV \in V$. Suppose that $iV = x$. Then, by hypothesis, $ix = x$, so $iV = ix$. So, by (23), $x = V$, a contradiction. So $iV \neq x$. Next, suppose that $iV = V$. Then V would have at most two members, but it in fact has at least three: x, Λ, and V. So $iV \neq V$. Suppose that $iV = \Lambda$. Then since $V \in iV$ we have $V \in \Lambda$. So, by (18), we have $V = \Lambda$ or $V \neq V$. But both alternatives are false. So $iV \neq \Lambda$. Finally, suppose that $iV = \{x, V, \Lambda\}$. Then since $\Lambda \in \{x, \Lambda, V\}$, we would have $\Lambda \in iV$. So, by (19), $\Lambda = V$ or $\Lambda = iV$. But both alternatives are again false. Thus, $iV \neq \{x, \Lambda, V\}$,

(d) $\Lambda \neq ix$ if $x \neq \Lambda$, $x \neq V$, $x \neq ix$.

Suppose that $\Lambda = ix$. Then $x \in \Lambda$ since $x \in ix$. But, by (20), $x = \Lambda$, contradicting the conditions of (d). So (d) is proved.

(e) $\Lambda \neq \{x, V, \Lambda\}$ if $x \neq \Lambda, x \neq V, x = ix.$

Suppose that $\Lambda = \{x, V, \Lambda\}$. Then $x \in \Lambda$, since $x \in \{x, V, \Lambda\}$. But then, by (20), $x = \Lambda$, contradicting the conditions of (e).

This completes the proof that the ranges of f_1 through f_4 are disjoint. It follows by a standard argument, which can be carried out in first-order logic, that f is one-to-one. We have thus shown that our function satisfies (12) and (14), and the inconsistency of Frege's system in the presence of (24) is immediate.

One might presume that the presence of contradictions in Frege's system and its repair are merely technical difficulties and do not vitiate his conception of sets. I find the situation much more serious, however. Not only do we lack a satisfactory repair for Frege's system, but the standard methods for avoiding the paradoxes either fail to preserve a significant part of set theory when applied to Frege's system or they are not compatible with Frege's conception of sets.

To see this, let us first note that Frege would *not* accept the standard approach, which takes the membership relation as primitive and introduces sets through axioms of set existence. To ensure their close tie to concepts, sets must be introduced through a second-level function that applies to concepts and yields objects as its values. This can be accomplished by taking class abstraction as a primitive function or by postulating the existence of such a function and defining class abstraction in terms of it. The difference between these two approaches is not important, but the former would seem to be more in the Fregean spirit. It is clear from Frege's discussions that class abstraction must be defined for all concepts and functions, that a separate hierarchy of types for classes is unacceptable, and that class abstracts must be treated as full-fledged designating singular terms. (These three points are made in Frege (1903a:254–256), but each is an application of more general doctrines concerning logic and its symbolism.) But this deprives Frege of the iterative approach to set theory as well as the (many-sorted) theory of types, von Neumann's distinction between sets and classes, and Quine's distinction between virtual and real classes. Thus, practically none of the major methods for dealing with the paradoxes developed since his time would be open to him.

Since Frege retained axiom Va, which we can put in the form

$$(x)(Fx \equiv Gx) \supset \hat{x}Fx = \hat{y}Gy,$$

class abstraction is to be regarded as a second-level function, but since (Vb) must be rejected, it cannot be a one-to-one function. Thus, there will be some concepts that apply to different objects and yet have the same extensions! In Frege's words, "this simply does away with extensions of concepts in the received sense of the term" (Frege 1903a:260–261). Frege thought that the modification in the notion of extension required was relatively minor: Two concepts with the same extension would apply to exactly the same objects, with the possible exception of the extension itself. He enshrined this conjecture in his repaired axiom V'b, which we have seen leads to a new contraduction. What he was unaware of was that *no finite number of exceptions is likely to yield a satisfactory repair*. A finite exception approach would use an axiom of the form

$$\hat{x}Fx = \hat{y}Gy \supset (x)(x \neq a_1 \ldots x \neq a_n \supset (Fx \equiv Gx)).$$

But this will lead, by an argument similar to that given earlier, to a new contradiction provided that there is a one-to-one function none of whose values is $a_1, a_2, \ldots, a_n$. Since such functions are readily available in any infinite domain, a system of this type can have no standard models in infinite domains. More generally, if we try to single out the exceptions via a condition Φ and use an axiom of the form

$$\hat{x}Fx = \hat{y}Gy \supset (x)(\Phi x \supset (Fx \equiv Gx)),$$

we must be wary of one-to-one functions whose values fail to satisfy the condition Φ. Thus, Frege's concept of sets and logic not only pushed him toward the Russell paradox but also seems to preclude a workable response to it.

The Circularity Objection

At the beginning of the century, Hilbert and Poincaré argued that Frege's reduction of number theory to logic is circular because it makes use of or presupposes number-theoretic methods. One form of this objection is the claim that inductive definitions and mathematical induction are needed to specify formal systems for logic and number theory and to prove that the latter has a translation in the former. But a Fregean always could retort that such procedures can be replaced by their logical (or set-theoretic) correlates. For example, in place of the inductive definition of *theorem,* one can use the explicit definition:

> *S* is a theorem just in case *S* belongs to every class that contains the axioms and is closed under the rules of inference.

This is an exact analog of Frege's definition of the class of natural numbers. But Poincaré produced a more subtle form of the circularity objection in response to this move (Poincaré:468–469). I will call this the *two-definitions objection,* and it runs as follows: Now we have two definitions of *theorem* (or *number* or *axiom* or etc.), an inductive one and an explicit set-theoretic one—but *induction* is needed to prove them coextensive! The following nice formulation of the two-definitions objection has been given by Charles Parsons, who credits it to Papert (Parsons:199). Within set theory we can define "*Nx*" (or the Frege numbers) by:

$$Nx \equiv (w)(0 \in w \cdot w \text{ is closed under successor} \supset x \in w).$$

Then in set theory we can prove

(1) $N0$
(2) $(x)(Nx \supset NSx)$

and by *modus ponens* and universal instantiation we can show $NS0,$ $NSS0,$ $NSSS0,$ for example. However, to prove that this process can be extended to show that every member of the inductively defined sequence $0, S0, SS0, \ldots$ is a Frege number we must apply induction in the metalanguage. Since to show that the two definitions are coextensive we must show that every inductively defined number is a Frege number and use unreduced induction to this end, the reduction of intuitive numbers to Frege numbers is circular.

No doubt Frege could respond with a set-theoretic replacement for induction at the metalinguistic level. But Poincaré could raise the same objection to that, and the subsequent regress would leave Frege forever unable to satisfy Poincaré's doubts. The burden here is also on Frege. For Poincaré had no doubts that the definitions *are* coextensive; he claimed only that this cannot be shown without an ultimate appeal to an unreduced induction principle:

> The two definitions are not identical; they are doubtless equivalent, but only in virtue of a synthetic judgment a priori; we cannot pass from one to the other by a purely logical procedure. [Poincaré:469]

Mark Steiner, in an extensive discussion of Parsons, argues that the admission that the two definitions are equivalent is enough to allow a logicist to avoid the objection. According to Steiner, logicism need only claim that arithmetical knowledge *can* be acquired via logic (or set theory). Poincaré's objection does not phase this claim. For, setting aside problems about Frege's logic as irrelevant to the current issue, certainly beliefs derived from the axioms of logic would be justified; and given the equivalence of the two definitions, our arithmetical beliefs can be so derived. To be sure, people who learn arithmetic via logic will lack the metatheoretic knowledge that the intuitive numbers *are* the Frege numbers. But because they will not have the conceptual apparatus to formulate that *nonarithmetical* claim, this should not count against logicism (Steiner:28–32).

Steiner argues correctly against Parsons and Poincaré in favor of logicism *as he construes it*. However, Frege (and historical logicism generally) was wedded to a stronger thesis than that considered by Steiner. Frege claimed, at least in his earlier writings, that arithmetic *is* a branch of logic and that numbers *are* logical objects. It is to these claims that Poincaré's objection is directed, and if he is correct, *their demonstration* requires the use of an induction principle that cannot be based upon logic. The situation is perplexing because it seems that if we are convinced of Poincaré's claims, then we should be convinced that arithmetical induction can be grounded logically although metalinguistic induction cannot! We can perhaps resolve this perplexity by applying to Poincaré the view on alternative reductions which I credited to Frege. This would interpret Poincaré as arguing that although arithmetic *has a model* in logic, this in itself is without epistemological significance. For the two-definitions objection shows that some form of mathematical induction ultimately must be taken as a primitive nonlogical principle.

This debate can be related to Frege's views on definitions, especially to those of his 1914 essay. You will recall that Frege distinguished two kinds of previously used expressions for which definitions are proposed. An expression of the first kind has a clear meaning and a proposed definition for it must be judged according to whether it captures this meaning. It will if and only if the identity between the *definiens* and *definiendum* is self-evidently true. Correct statements of this sort are, according to Frege, not really definitions and should be accorded the status of axioms. An expression of the second kind has no clear meaning, and it is appropriate to define it anew. The new definition, however, should be viewed as an explication, and the question of the relationship between the old and new meanings of the expression it defines should be bypassed.

Now where does Frege's definition of number fit into this scheme? Poincaré might argue that Frege's definition is, strictly speaking, an axiom. The expression "natural number" already has a clear meaning, so he could claim, and the correctness of Frege's "definition" is self-evident but also synthetic a priori.

Frege would respond that our previous conception of the natural numbers was not an adequate one. When he first proposed his analysis, neither number theory nor the natural numbers had been characterized precisely. Thus, it would have been proper for him to view his definition as an explication of the concept of number. But what about today, when we have the Peano axioms and extensive studies of recursive definitions and inductive reasoning? A Fregean could still argue that the intended interpretation of axiomatic number theory makes use of the notion of an arbitrary finite iteration, and *that notion* is what Frege's definition explicates.

[*223*]

Thus, we are led to a stalemate. The pro-Fregean will fall back upon the notion of a finite set and claim that it is clear and clarifies the notion of natural number. The anti-Fregeans will claim that just the opposite is the case: The notion of a finite set depends for its clarity upon that of a natural number. (Cf. Steiner:77.) In the end it seems that we are forced to choose between conflicting intuitions about which putative foundation is more fundamental. I think that this unsatisfactory situation might be resolved by arguing that in some sense both parties are dealing with the same basic mathematical structure, but simply approaching it via different descriptive means. That they are in some sense dealing with the same mathematical structure is evident from the fact that each of their theories is isomorphic to a subset of the other's. Thus, the situation is analogous to that of two structures *A* and *B*, where *A* is an infinite sequence consisting of a circle with an inscribed square, which in turn contains an inscribed circle, and so on, and *B* is the same type of sequence except that it begins with a square. It seems rather silly to ask whether *A* is more fundamental than *B* because not only does the choice between a circle and a square as more basic (or outer) seem completely arbitrary in this context, but also in a graphic sense the two sequences are contained in each other. I wish to suggest that, by thinking of the problem whether numbers or sets are more fundamental along the lines of our two geometrical sequences, we might come to understand how and why our conflicting intuitions here simply amount to differences of taste.

Logic vs. Set Theory

A much more telling objection to Frege's reduction is that the "logic" to which he reduced number theory is really a rather strong set theory. We have seen in discussing Frege's conception of sets how and why he thought that *his sets,* at least, were logical objects. In pursuing this objection we should think of the reduction of number theory as being carried out in at least a reasonable approximation to Fregean set theory (if a consistent one is possible). Even so, we can conclude from the nature of number theory important properties about the power of this (imaginary) set theory, and thus bring out important differences between what everyone takes to be at least the core of logic—first-order logic and Fregean set theory. As has been remarked before by Benacerraf, Parsons, and Steiner (Steiner:71–72): (1) the truths of first-order logic hold in all nonempty universes, but a set theory capable of yielding numbers as sets will require an infinite universe; (2) the axioms of first-order logic are in a sense obvious but no such claim can be made any longer for Fregean set theory (a point granted by Frege himself (1903a:253)); (3) first-order logic is complete, while a set theory adequate for number theory must be incomplete; and (4) the concept of set, especially the Fregean concept of set, is far more complicated and less understood than the notions underpinning first-order logic. This has been demonstrated by the difficulty we have had in arriving at a satisfactory set of axioms for set theory, the doubts cast by the Skolem Paradox on our understanding of the distinctions between finite and infinite sets and countable and uncountable sets, and the difficulty we now face in assigning a truth-value to the continuum hypothesis.

It is generally accorded that an epistemologically adequate criterion of logical truth is lacking. Despite this, the considerations just enunciated clearly show that on the grounds of complexity and lack of certainty, set theory would not belong in the same epistemological category as first-order logic (if any satisfactory concept of logical truth is to be had).

In light of this, how should we evaluate Frege's attempt to prove that arithmetic is analytic? Obviously, he operated with a stronger notion of analyticity than did Kant. Frege attributed this to his use of more sophisticated forms of definition than were available to Kant (Frege 1884:sec. 88). It is surprising to note, however, that Frege's notion is even broader than that which contemporary philosophers often take themselves to have obtained from him. In particular, today we usually assume that an identity

$$s = t$$

is analytic just in case the terms s and t are synonymous. But this will hardly fit with the following passage from Frege:

I distinguish from the *denotation* of a name its *sense*. "2^2" and "$2 + 2$" do not have the same *sense,* nor do "$2^2 = 4$" and "$2 + 2 = 4$" have the same *sense.* [Frege 1893:sec. 2]

For "$2 + 2 = 4$," "$2^2 = 4$," and "$2^2 = 2 + 2$" are all analytic according to Frege, and yet "2^2," "$2 + 2$," and "4," by virtue of having different senses, are nonsynonymous.

These cases are quite striking because one wants to say: Surely "$2 + 2 = 4$," for instance, must arise from "$4 = 4$" by substituting terms that are related to each other through definitions; and Frege himself claimed that a *definiens* and a *definiendum* are synonymous. And, clearly, synonymy is transitive. So how can you explain the non-synonymy of "$2 + 2$" and "4"?

The answer is to be found in a fuller analysis of Frege's derivation of "$2 + 2 = 4$" which depends not *only* upon identities of the form "$a = a$" and definitions but *also* makes use of the recursion equations for addition:

(1) $x + s(y) = s(x + y)$; (2) $x + 0 = x$	
$2 + 2 = 2 + 2$	law of identity
$2 = s(1)$	definition
$2 + 2 = 2 + s(1)$	sub $=$
$2 + s(1) = s(2 + 1)$	(1)
$1 = s(0)$	definition
$2 + 2 = s(2 + s(0))$	sub $=$
$2 + s(0) = s(2 + 0)$	(1)
$2 + 0 = 2$	(2)
$2 + 2 = ss(2)$	sub $=$
$s(2) = 3$	definition
$s(3) = 4$	definition
$2 + 2 = s(3)$	sub $=$
$2 + 2 = 4$	sub $=$

The recursion equations for addition are in turn derived from the definition of addition through *second-order* logic. More specifically, if we define "Σ" by

$\Sigma abc \equiv (F)[(x)Fx0x \cdot (x)(y)(z)(Fxyz \supset Fxs(y)s(z)) \supset Fabc]$,

then we can prove (using second-order logic)

$$(x)\Sigma \, x0x \cdot (x)(y)(z)(\Sigma xyz \supset \Sigma xs(y)s(z)).$$

Then by defining "$x + y$" as "$(\imath z)\Sigma xyz$," we can obtain the usual recursion equations. (Ultimately, these derivations also depend upon laws governing successor and zero, which Frege obtained through set theory.)

Much of the appeal of the notion of analyticity derives from the idea that an analytical truth can be established via an analysis of the terms used to express it (without appealing to "extraneous" principles). Frege's introduction of set theory clearly detracts from this conception of analyticity, but it seems questionable to me that even second-order logic should be used in grounding the concept. For instance, in establishing the truth of

> if John is an ancestor of both Bill and Peter, then one of
> them is an ancestor of the other provided that John and
> all his descendants have one child each,

we must go through a fairly complex chain of inferences and appeal to strong principles of "ancestor induction." Yet, on Frege's view this truth would count as analytic because it can be proved using second-order logic and an appropriate definition of the ancestor relation.

Ironically, the inferences needed to prove this truth are strictly analogous to the inductive arguments used in proving that *less than or equal* is a connected relation among the natural numbers. Thus, we have come full circle. Frege started in 1879 with the question of whether mathematical reasoning, in particular inference via mathematical induction, is a species of logical reasoning. His affirmative answer to this question was founded upon the reduction of mathematical induction to an apparently much more general form of reasoning. But we can see now that whether we count this form of reasoning as a type of logical inference may depend upon our intuitions concerning inductive reasoning itself.

Gödel's incompleteness theorem poses a problem for all programs with the goal of reducing mathematics to logic. Logic ought to have a complete and effective set of axioms—since our paradigm, first-order logic, does—but number theory cannot. There are a number of responses to this. For instance, the reduction can be aimed at some "standard" mathematics, characterized more precisely as the consequences of a fixed set of axioms. Or one can argue against the emphasis on the completeness of logic. But the Gödel theorem delivers a hard blow to Frege's reductionist program.

Frege aimed to reduce all of arithmetic to logic. Although he never precisely characterized arithmetic, it is all arithmetical truth that he had in mind. Anything less would conflict with his Platonism. On the other hand, Frege insisted that his reduction be based upon effective proofs from a finite set of axioms (Frege 1884:secs. 5, 91). The Gödel theorem shows that he cannot meet these desiderata and fulfill his reductive program, no matter how liberal he is otherwise in delimiting logic.

The Problem of Multiple Reductions

In discussing Frege's definition of number and his reduction of number theory to set theory, we saw that it met all the conditions that any successful reduction of number theory should: it furnished a model for the Peano axioms and provided foundations for the theory of counting. The trouble is, as Paul Benacerraf has observed (Benacerraf), many other reductions also meet these conditions. In fact, suppose that in the context of set theory we take 0 and successor as *primitive* and governed by the axioms

(a) $(x)(0 \neq S(x))$
(b) $(x)(y)(S(x) = S(y) \supset x = y)$.

Then we can define the natural numbers, and the less than and equicardinality relations, all in the manner of Frege, and develop the Peano axioms and the theory of counting (Quine 1963:74–81). Thus, all that remains to complete the reduction is definitions of 0 and successor that enable us to derive the axioms (a) and (b). Frege gave us one definition, but Zermelo's

$$0 = \Lambda$$
$$S(x) = \{x\}$$

and von Neumann's

$$0 = \Lambda$$
$$S(x) = x \cup \{x\}$$

work just as well. So does

$$0 = \{\Lambda\} \cup \{\{\Lambda\}\}$$
$$S(x) = \{x\}$$

or infinitely many others obtained from, say, Zermelo's reduction by taking his number n as the new zero (and adjusting the successor relation so that zero is not a successor). Grounds for some choice exist in the context of some larger set-theoretic purpose. Zermelo's and von Neumann's definitions will not work in the theory of types, since each number will be of a different type, and von Neumann's is not suited to the theory of finite sets (Quine 1963:81–83). Yet even such considerations still leave us with too much latitude. For given any progression now serving as a reduction, a new candidate may be obtained by choosing as the new zero any member of the first progression and taking the members of the progression following the new zero as the other new numbers. (Charles Chihara (1980) has questioned the force of this argument.)

Now what has this to do with Frege? Benacerraf argues as follows. Frege maintained that numbers are classes. But if they are classes, then they must be particular classes. Which class, then, is the number two? Frege said that it was the class of all pairs. But why not identify it with $\{1\}$ (following Zermelo) or $\{0, 1\}$ (following von Neumann)? Of course, it can be identical with at most one of these, but which one? As we saw above, there are no mathematical or other "cognitive" considerations that will resolve these questions. But if the number two (or any other number) is not any particular class, then how can it or any other number be a class?

Benacerraf pushes this line of thought a step further (Benacerraf:69–70). Since *any progression* will serve to complete our account of numbers, it follows that numbers are not objects. For if they are objects, then each number must be a particular object in some particular progression of objects. But again, which object is the number two? Is it $\{0, 1\}$, or is it the third member of the sequence of stroke symbols: 1, 11, 111, 1111, ... ? Again there are no conclusive grounds for a decision. Benacerraf concluded that ''numbers could not be objects at all: for there is no more reason to identify any individual number with any one particular object than with any other (not already known to be a number)'' (Benacerraf:69).

This form of argument can be used to discredit many other mathematical reductions. For example, real numbers can be reduced to either Dedekind cuts of rationals or Cantorian series or Cauchy sequences or to variations on these themes. There are also alternative reductions of the ordinal numbers to sets, of functions to sets, and of ordered pairs to sets. The Benacerraf argument would have us conclude that ordered pairs are not sets and that the reals are not ''logical constructions'' from the rationals. More generally, if we have a theory of objects of kind A and alternative and ''equally correct'' reductions of this theory to theories of objects of kinds B and C which themselves are of kind D, then we are to conclude that the A's cannot be of kind D. The argument will be widely applicable. For it is to be expected that if one reduction of the A's to the D's is available, then others will be forthcoming.

(Whether the last claim can be rigorously proved to hold generally depends upon how ''reduction'' is defined. For example, if we say that the A's reduce to the D's just in case the theory of the A's has a model whose domain is the class of D, then alternative ''reductions'' to D are easily obtainable by using standard model theoretic techniques that ''permute'' the elements of a model in order to obtain a new one. On the other hand, given stricter (and more plausible) syntactic definitions of reduction, such general results must fail, as Richard Grandy has shown me.)

[*230*]

The force of Benacerraf's argument thus appears to be so general that we may fear that no reductions, or no interesting reductions at least, are ever possible. The response is not, I think, to question the validity of Benacerraf's argument but rather to take a closer look at the purpose of a reduction and our expectations from it.

Before turning to this, let us note that Benacerraf has not proved that numbers are not objects. What he appears to have proved is that if "*n*" is a numeral and "*a*" is the name of an object not already known to be a number, then we have no justification for asserting that "*n* = *a*" rather than some other identity, say, "*n* = *b*." Thus, we might be forced to conclude that numbers are not sets, nor tables and chairs, nor stroke symbols; yet *we are not forced to conclude that numbers are not numbers.* As we have seen earlier, the Fregean thesis that numbers are objects is best viewed as a thesis concerning the logical form of numerical statements and numerical terms. Benacerraf has done nothing to shake this thesis. Rather, his argument presupposes that something is an object only if it is identical with a member of a certain collection of things not usually taken to be numbers, such as sets, chairs, or tables. To assume without further argument that every object belongs to this collection is to beg the question.

A new twist can be obtained by starting with the premise that classes are the only abstract entities which we need to countenance, since they can be made to fill the roles that other abstract objects play (Quine 1960:266–270). One could argue, then, that classes are the only abstract objects; and thus numbers are not objects, because they are abstract but not classes. (Cf. Kitcher 1978.) The trouble is that the original premise already runs afoul of Benacerraf's argument. For the existence of alternative reductions of other abstract objects (besides numbers) to classes will undermine the claim that classes are the only abstract objects. So much for the argument that numbers are not objects.

Turning now to the problem of reductions, it is clear that we want to accept some reductions and to do so must parry the Benacerraf argument. Ironically, Frege's later writings on definitions point the way to do so. In the case of the reduction of numbers to sets, for instance, we bypass the question of whether numbers are sets and simply propose that set theory with number theory added to it be replaced by set theory alone. The grounds for doing so are those of theoretical simplicity and ontological economy. If we do *not* make the replacement, then we will be left with a mathematics that is committed to both sets and numbers and is unable to give a nonarbitrary answer to the question of whether numbers are sets. If we make the replacement, we will avoid the (possible) extraontic commitments and the unanswerable questions. (It should be pointed out that Benacerraf himself conceded that his argument does not apply to ontological reductions construed as explications in this way. He did not seem aware that Frege, at least the later Frege, had this escape route.) When our conception of the entities to be eliminated is faulty, then we have another motive for this approach to reduction. For it lets us jettison our old faulty theory without a backward glance. Frege, we can be sure, would urge that approach to his contemporaries' theories of number and counting.

Frege does not, I think, escape unscathed from a point emanating from Benacerraf's argumentation. Our knowledge of numbers, sets, and other mathematical objects can be communicated only via assertions that purport to be true of them. Now let us be given as many true sentences of number theory as you like (axiomatized or not). These can be construed alternatively as a theory about the Zermelo numbers, the von Neumann numbers, and so on. Set theory is subject also to alternative interpretations, since sets can be construed in terms of categories. And, in general, any theory that has one model will have infinitely many isomorphic ones. But if our talk of sets, numbers, and other mathematical entities is thus construable in indefinitely many alternative ways, how can we claim that our mathematical terms have fixed references?

Perhaps one might reply that we can fix the reference of a term by "grasping" or "apprehending" a definite Fregean sense and associating it with the term, which will in turn determine the reference of the term. Of course, this puts a heavy burden on Frege's notion of the apprehension of a sense and retreats to a completely nonexplanatory account of reference to mathematical entities. If it were Frege's response, then it certainly would count against his theory. I am not sure that it is his response. For when he encountered the systematic referential indeterminacy of class abstracts arising from his introduction of them by means of his axiom V, he attempted to resolve this indeterminacy by stipulation rather than by retreating to the doctrine of an intended reference. But, as we saw, his valiant efforts were of no avail. How Frege would react to the more complex indeterminacy that modern results have revealed, I cannot venture to say. No clear answer is evident in his writings.

Concluding Remarks

Let us now try to assess the accomplishments and defects of Frege's work and its impact upon the philosophy of mathematics. We should begin by giving him the credit he deserves for his many technical advances, since we are apt to lose sight of them in the face of the criticisms of his views discussed above. To repeat an earlier point: The essentials of his analysis of the logical form of numerical judgments and numerical terms are still recognized today. They have been disassociated from his theory of concepts and objects. But part of their value is that they do not stand or fall with that theory. The same remarks hold for Frege's theory of counting, his analysis of inductive reasoning in terms of second-order logic, and his definition of numbers in terms of sets. All of these stand as mathematical and logical results to Frege's credit, but we should note that the philosophical consumer need not buy them as a single package. For example, one philosophically suspicious of sets can reject the definitions of numbers as sets while assimilating Frege's other results. Even a constructivist can make use of a suitably modified version of Frege's results concerning counting and the logical form of numerical judgments.

Frege also has done much to advance the cause of Platonism as a philosophy of mathematics. We saw in discussing psychologism, formalism, and deductivism that Frege's and subsequent critiques have failed to unearth absolutely fatal difficulties in these views. Nevertheless, each of these views is difficult to defend in the face of a well-wrought Platonism, since the latter offers a particularly simple and convincing account of language, practice, and place of mathematics in the sciences. The serious philosophical difficulties with Frege's account did not surface immediately, unlike the problems with his set theory.

The reduction of arithmetic to "logic" would seem to have removed the greatest obstacle to the acceptance of Platonism since it appeared to provide an epistemology for arithmetic. As a matter of history, many of the logicists of the first one-third of this century did not follow Frege's lead in embracing ontological and epistemological Platonism: Russell proclaimed the no-class theory, and the logical positivists opted for conventionalism in mathematics. Yet today Platonism is one of the few serious contenders in the philosophy of mathematics. Undoubtedly, this is due in part to the failures of other approaches and to the influence of Gödel and Quine (both Platonist admirers of Frege), but much is due also to the rediscovery of Frege's penetrating criticisms of his contemporaries and his persuasive apologies for his own views.

Yet Frege did not provide all the answers. Since it seems impossible to develop a truly "logical" conception of classes that will furnish a consistent and adequate foundation for mathematics, the Fregean attempt to found a (partial) epistemology for mathematics through logic seems destined to fail. Arithmetic and set theory must be recognized as mathematical structures and an epistemology for them developed. Along with this we must solve the other problem that the study of Frege has discovered, the problem of reference to mathematical objects. It would not be surprising if the solutions to these two problems came together. For if we can determine how we know mathematical objects and what we can know of them, then we should have some idea of how we relate to them. That, in turn, could provide a clue to how we refer to them.

Bibliography

Bartlett, J. M. 1961. *Funktion und Gegenstand*. Dissertation, Ludwig-Maximilians-Universität. Munich: M. Weiss.

Benacerraf, P. 1965. "What Numbers Could Not Be." *Philosophical Review,* 74:47–73.

Benacerraf, P., and H. Putnam, eds. 1964. *Philosophy of Mathematics*. Englewood Cliffs, N.J.: Prentice-Hall.

Bernays, P. 1942. Review of "Ein unbekannter Brief von Gottlob Frege über Hilberts erste Vorlesung über die Grundlagen der Geometrie," ed. by M. Steck. *Journal of Symbolic Logic,* 7:92–93.

Bishop, E. 1967. *Foundations of Constructive Analysis*. New York: McGraw-Hill.

Boolos, G. 1971. "The Iterative Conception of Set." *Journal of Philosophy,* 68:215–230.

———. 1975. "On Second-Order Logic." *Journal of Philosophy,* 72:509–527.

Bostock, D. 1974. *Logic and Arithmetic*. Oxford: Clarendon Press.

Brouwer, L. E. J. 1913. "Intuitionism and Formalism," trans. by A. Dresden. *Bulletin of The American Mathematical Society,* 20:81–96. Reprinted in Benacerraf and Putnam.

Bynum, T., trans./ed. 1972. *Gottlob Frege: Conceptual Notation and Related Articles*. Oxford: Clarendon Press.

Chihara, C. 1980. "Ramsey's Theory of Types: A Return to Fregean Sources," in *Prospects for Pragmatism: Essays in Memory of F. P. Ramsey,* ed. by H. Mellor. Cambridge: Cambridge University Press.

Church, A. 1951. "A Formulation of the Logic of Sense and Denotation," in *Structure, Method and Meaning,* ed. by P. Henle, M. Kallen, and S. Langer. New York: Liberal Arts Press.

Currie, G. 1978. "Frege's Realism." *Inquiry,* 21:218–221.

Curry, H. B. 1951. *Outlines of a Formalist Philosophy of Mathematics*. Amsterdam: North-Holland.

BIBLIOGRAPHY

Dedekind, R. 1888. *Was sind und was sollen die Zahlen?* Brunswick.

Detlefsen, M. 1979. "On Interpreting Gödel's Second Theorem." *Journal of Philosophical Logic*, 8:297–313.

Dummett, M. 1973. *Frege: Philosophy of Language.* New York: Harper & Row.

——. 1976. "Frege as a Realist." *Inquiry*, 19:455–492.

Frege, G. 1873. *Über eine geometrische Darstellung der imaginären Gebilde in der Ebene.* Doctoral dissertation, Göttingen. Jena: A. Neuenhann.

——. 1879a. "Anwendungen der Begriffsschrift." *Sitzungsberichte der Jenaischen Gesellschaft für Medicin und Naturwissenschaft, Jenaische Zeitschrift für Naturwissenschaft,* 13:29–33. Translated in Bynum.

——. 1879b. *Begriffsschrift, eine der arithmetischen nachgebildete Formelsprache des reinen Denkens.* Halle: L. Nebert. Translated in Van Heijenoort and in Bynum.

——. 1884. *Die Grundlagen der Arithmetik.* Reprinted and translated by J. Austin in *The Foundations of Arithmetic.* Oxford: Blackwell and Mott, 1950.

——. 1885. "Über formale Theorien der Arithmetik." *Sitzungsberichte der Jenaischen Gesellschaft für Medicin und Naturwissenschaft, Jenaische Zeitschrift für Naturwissenschaft,* 19:94–104. Translated in Kluge.

——. 1891. *Funktion und Begriff.* Jena: H. Pohle. Translated in Geach and Black.

——. 1892a. "Über Sinn und Bedeutung." *Zeitschrift für Philosophie und philosophische Kritik,* 100:25–50. Translated in Geach and Black.

——. 1892b. "Über Begriff und Gegenstand." *Vierteljahrsschrift für wissenschaftliche Philosophie,* 16:192–205. Translated in Geach and Black.

——. 1893. *Grundgesetze der Arithmetik,* Vol. I. Jena: H. Pohle. Translated in part by M. Furth in *The Basic Laws of Arithmetic: Exposition of the System.* Los Angeles: University of California Press.

——. 1894. Review of E. Husserl, *Philosophie der Arithmetik, Zeitschrift für Philosophie und philosophische Kritik,* 103:313–332. Translated in part in Geach and Black.

——. 1895. "Kritische Beleuchtung einiger Punkte in E. Schröders Vorlesungen über die Algebra der Logik." *Archiv für systematische Philosophie,* 1:433–456. Translated in Geach and Black.

——. 1896. "Lettera del sig. G. Frege all'editore." *Rivista di Matematica* (Revue de Mathematiques), 6:53–59.

——. 1903a. *Grundgesetze der Arithmetik,* Vol. II. Jena: H. Pohle. Translated in part in Geach and Black.

——. 1903b. "Über die Grundlagen der Geometrie." *Jahresberichte der Deutschen Mathematiker-Vereinigung,* 12:319–324, 368–375. Translated in Kluge.

——. 1904. "Was ist eine Funktion?" in *Festschrift Ludgwig Boltzmann gewidmet zum sechzigsten Geburtstage, 20. Februar 1904.* Leipzig: A Barth. Translated in Geach and Black.

———. 1906a. "Über die Grundlagen der Geometrie, I, II, III." *Jahresberichte der Deutschen Mathematiker-Vereinigung,* 15:293–309, 377–403, 423–430. Translated in Kluge.

———. 1906b. "Antwort auf die Ferienplauderei des Herrn Thomae." *Jahresberichte der Deutschen Mathematiker-Vereinigung,* 15:586–590. Translated in Kluge.

———. 1908. "Die Unmöglichkeit der Thomaeschen formalen Arithmetik aufs neue nachgewiesen." *Jahresberichte der Deutschen Mathematiker-Vereinigung,* 17:52–55. Translated in Kluge.

———. 1912. "Remarks on P. Jourdain, 'The Development of the Theories of Mathematical Logic and the Principles of Mathematics.'" *Quarterly Journal of Pure and Applied Mathematics,* 43:237–269.

———. 1918. "Der Gedanke: eine logische Untersuchung." *Berträge zur Philosophie des deutschen Idealismus,* 58–77. Translated by A. Quinton and M. Quinton in *Mind,* 65 (1956):289–311.

———. 1923. "Logische Untersuchungen; Dritter Teil: Gedankengefüge." *Beiträge zur Philosophie des deutschen Idealismus,* 3:35–51. Translated by R. Stoothoff in *Mind,* 72 (1963):1–17.

———. 1940. "Ein unbekannter Brief von Gottlob Frege über Hilberts erste Vorlesung über die Grundlagen der Geometrie," ed. by M. Steck. *Sitzungsberichte der Heidelberger Akademie der Wissenschaften: Mathematisch-naturwissenschaftliche Klasse, Jahrgang 1940.* Translated in Kluge.

———. 1941. "Unbekannte Briefe Freges über die Grundlagen der Geometrie und Antwortbrief Hilberts an Frege," ed. by M. Steck. *Sitzungsberichte der Heidelberger Akademie der Wissenschaften: Mathematisch-naturwissenschaftliche Klasse, Jahrgang 1941.* Translated in Kluge.

———. 1967. *Kleine Schriften,* ed. by I. Angelelli. Darmstadt: Wissenschaftliche Buchgesellschaft. (Contains reprints of most of Frege's published articles.)

———. 1969. *Nachgelassene Schriften,* ed. by H. Hermes, F. Kambartel, and F. Kaulbach. Hamburg: Felix Meiner.

Freudenthal, H. 1962. "The Main Trends in the Foundations of Geometry in the 19th Century," in *Logic, Methodology and Philosophy of Science,* ed. by E. Nagel, P. Suppes, A. Tarski. Stanford: Stanford University Press.

Geach, P. T. 1956. "On Frege's Way Out." *Mind,* 65:408–409.

Geach, P., and Black, M. 1952. *Translations from the Philosophical Writings of Gottlob Frege.* Oxford: Blackwell.

Goodman, N. 1966. *The Structure of Appearance,* 2d ed. Indianapolis: Bobbs-Merrill.

Henkin, L. 1950. "Completeness in the Theory of Types." *Journal of Symbolic Logic,* 15:81–91.

BIBLIOGRAPHY

Heyting, A. 1956. *Intuitionism*. Amsterdam: North-Holland.

Hilbert, D. 1899. *Die Grundlagen der Geometrie*. Leipzig.

——. 1900a. "Mathematisches Problem." *Nachrichten von der königlichen Gesellschaft der Wissenschaften zu Gottingen:* 253–297.

——. 1900b. "Über den Zahlbegriff." *Jahresbericht der Deutschen Mathematiker-Vereinigung*, 8:180–194.

——. 1904. "Über die Grundlagen der Logik und der Arithmetik." *Verhandlungen des Dritten Internationalen Mathematiker-Kongresses in Heidelberg vom 8. bis 13. August 1904*. Leibzig: Teubner. Translated in Van Heijenoort.

——. 1922. "Neubegründung der Mathematik." in *Abhandlungen aus dem Mathematischen Seminar der Hamburgischen Universität*, 1:157–177. Pagination cited from D. Hilbert, *Gesammelte Abhandlungen*, Bd. III. Berlin: Springer, 1935.

——. 1925. "Über das Unendliche." *Mathematische Annalen*, 95:161–190. Translated in Benacerraf and Putnam and in Van Heijenoort.

——. 1927. "Die Grundlagen der Mathematik." *Abhandlungen aus dem Mathematischen Seminar der Hamburgischen Universität*, 6:65–85. Translated in Van Heijenoort.

Hilbert, D., and P. Bernays. 1938 and 1939. *Die Grundlagen der Mathematik*, 2 vols. Berlin: Springer.

Husserl, E. 1970. *Philosophie der Arithmetik (Gesammelte werke*, Vol. XII), ed. by L. Eley. The Hague: M. Nijhoff. (First published in 1891.)

Kessler, G. 1978. "Mathematics and Modality." *Noûs*, 12:421–441.

Kitcher, P. 1976. "Hilbert's Epistemology." *Philosophy of Science*, 43:99–115.

——. 1978. "The Plight of the Platonist." *Noûs*, 12:119–136.

——. 1979. "Frege's Epistemology." *Philosophical Review*, 88:235–262.

Klemke, E. D., ed. 1968. *Essays on Frege*. Urbana, Ill.: University of Illinois Press.

Kluge, E.-H., trans./ed. 1971. *Gottlob Frege: On the Foundations of Geometry and Formal Theories of Arithmetic*. New Haven: Yale University Press.

Kreisel, G. 1958. "Hilbert's Programme." *Dialectica*, 12:346–372. Reprinted in Benacerraf and Putnam.

Locke, J. 1689. *An Essay Concerning Human Understanding*. New York: Dover, 1959.

Lotze, H. 1880. *Logik*, 2d ed. Leipzig.

Mendelson, E. 1964. *Introduction to Mathematical Logic*. Princeton, N.J.: D. Van Nostrand.

Mill, J. S. 1843. *A System of Logic*. New York: Longmans, Green and Co., 1936.

Parsons, C. 1965. "Frege's Theory of Number," in *Philosophy in America*, ed. by M. Black. Ithaca, N.Y.: Cornell University Press.

[238]

Peirce, C. 1902. "The Essence of Mathematics," in *Peirce: Essays in the Philosophy of Science*, ed. by V. Tomas. Indianapolis: Bobbs-Merrill (1957).

Poincaré, H. 1913. *The Foundations of Science*, trans. by G. B. Halsted. New York: The Science Press.

Putnam, H. 1967a. "The Thesis That Mathematics Is Logic," in *Bertrand Russell: Philosopher of the Century*, ed. by R. Schoenman. Boston: Little, Brown. Reprinted in *Mathematics, Matter and Method*.

——. 1967b. "Mathematics without Foundations." *Journal of Philosophy*, 64:5–22. Reprinted in *Mathematics, Matter and Method*.

——. 1975. "What Is Mathematical Truth?" in *Mathematics, Matter and Method*. Cambridge, England: Cambridge University Press.

Quine, W. V. 1936. "Truth by Convention," in *Philosophical Essays for A. N. Whitehead*, ed. by O. H. Lee. New York: Longmans. Reprinted in Benacerraf and Putnam.

——. 1951. "Two Dogmas of Empiricism." *Philosophical Review*, 60:20–46. Reprinted in Benacerraf and Putnam.

——. 1955. "On Frege's Way Out." *Mind*, 64:145–159.

——. 1960. *Word and Object*. Cambridge, Mass.: M.I.T. Press.

——. 1962. "Carnap and Logical Truth," in *Logic and Language: Studies Dedicated to Professor Rudolf Carnap on the Occasion of His Seventieth Birthday*. Dordrecht: D. Reidel.

——. 1963. *Set Theory and Its Logic*. Cambridge, Mass.: Harvard University Press.

Reid, C. 1970. *Hilbert*. Berlin: Springer.

Resnik, M. 1963. "Frege's Methodology." Ph.D. dissertation, Harvard University.

——. 1965. "Frege's Theory of Incomplete Entities." *Philosophy of Science*, 32:329–341.

——. 1967. "The Role of the Context Principle in Frege's Philosophy." *Philosophy and Phenomenological Research*, 27:356–365.

——. 1974. "The Philosophical Significance of Consistency Proofs." *Journal of Philosophical Logic*, 3:133–147.

——. 1976. "Frege's Context Principle Revisited," in *Studies on Frege*, Vol. III, ed. by M. Schirn. Stuttgart-Bad Cannstatt: E. Frommann.

Russell, B. 1937. *The Principles of Mathematics*, 2d ed. London: Allen & Unwin.

Sheppard, P. 1973. "A Finite Arithmetic." *Journal of Symbolic Logic*, 38:232–248.

Sigwart, C. 1895. *Logik*. 2 vols. Trans. by H. Dendy. New York: Macmillan.

Sluga, H. D. 1975. "Frege and the Rise of Analytical Philosophy." *Inquiry*, 18:471–484.

——. 1976. "Frege as a Rationalist," in *Studies on Frege*, Vol. 1, ed. by M. Schirn. Stuttgart-Bad Cannstatt: F. Frommann.

BIBLIOGRAPHY

——. 1977. "Frege's Alleged Realism." *Inquiry,* 20:227-242.
Smorynski, C. 1977. "The Incompleteness Theorems," in *Handbook of Mathematical Logic,* ed. by Jon Barwise. Amsterdam: North-Holland.
Snapper, J. 1974. "Contextual Definition: What Frege Might Have Meant but Probably Didn't." *Noûs,* 8:259-272.
Sobocinski, B. 1949. "L'Analyse de l'antinomie russellienne par Leśniewski: IV. La correction de Frege." *Methodos,* 1:220-228.
Steiner, M. 1975. *Mathematical Knowledge.* Ithaca, N.Y.: Cornell University Press.
Suppes, P. 1957. *Introduction to Logic.* Princeton, N.J.: D. Van Nostrand.
Thiel, C. 1965. *Sinn und Bedeutung in der Logik Gottlob Freges.* Meisenheim am Glan: Anton Hain.
Thomae, J. 1906. "Gedankenloser Denker, eine Ferienplauderei." *Jahresberichte der Deutschen Mathematiker-Vereinigung,* 15:434-438. Translated in Kluge.
Van Heijenoort, J., ed. 1967. *From Frege to Gödel.* Cambridge, Mass.: Harvard University Press.
Von Neumann, J. 1931. "The Formalist Foundations of Mathematics." *Erkenntnis:* 91-121. Reprinted in Benacerraf and Putnam.

Index

[241]

FREGE AND THE PHILOSOPHY
OF MATHEMATICS

Designed by Richard E. Rosenbaum.
Composed by The Composing Room of Michigan, Inc.
in 10 point Times Roman V.I.P., 3 points leaded,
with display lines in Times Roman.
Printed offset by Thomson/Shore, Inc. on
Warren's Number 66 Antique Offset, 50 pound basis.
Bound by John H. Dekker & Sons, Inc.
in Holliston book cloth
and stamped in Kurz-Hastings foil.

Library of Congress Cataloging in Publication Data

Resnik, Michael D
   Frege and the philosophy of mathematics.

   Bibliography: p.
   Includes index.
   1. Mathematics—Philosophy. 2. Frege, Gottlob,
1848-1925. I. Title.
QA8.4.R47   510′.1   80-11120
ISBN 0-8014-1293-5